Worldwide Distributor:

RUSH Publications and Educational Consultancy, LLC
1901 60th Place, L7432, Bradenton
FL 34203
United States

All inquiries, suggestions and comments should be addressed to:

ARMA Publishing and Trade Ltd.
Ferhat S Bozbey A Daire 1 Rumeli Hisarustu
Istanbul, 34340
TURKEY

e-mail	:	meylani@superonline.com
phone	:	+90 212 263 5742, +90 212 263 5743
mobile	:	+90 533 363 5287, +90 506 541 8729
fax	:	+90 533 982 6740

ISBN : 0974886882

Author : Ruşen MEYLANİ
Cover : Pınar ERKORKMAZ; **e-mail:** mail@pinare.com

All graphics and text based on TI 83 – 84 have been used with the permission granted by Texas Instruments.

From: Bassuk, Larry <l-bassuk@ti.com>
Sent: Thursday, February 27, 2002 16:23 PM
To: Meylani, Rusen <meylani@superonline.com>; Foster, Herbert <h-foster@ti.com>; Vidori, Erdel <e-vidori@ti.com>
RE: USAGE OF TI 83+ FACILITIES IN MY SAT II MATH BOOKS
Rusen Meylani,
Again we thank you for your interest in the calculators made by Texas Instruments.
Texas Instruments is pleased to grant you permission to copy graphical representations of our calculators and to copy graphics and text that describes the use of our calculators for use in the two books you mention in your e-mail below.
We ask that you provide the following credit for each representation of our calculators and the same credit, in a way that does not interrupt the flow of the book, for the copied graphics and text:
Courtesy Texas Instruments
Regards,
Larry Bassuk
Copyright Counsel
972-917-5458

-----Original Message-----
From: Bassuk, Larry
Sent: Thursday, February 21, 2002 9:14 AM
To: 'Rusen Meylani'; Foster, Herbert
Subject: RE: USAGE OF TI 83+ FACILITIES IN MY SAT II MATH BOOKS
We thank you for your interest in TI calculators.
I am copying this message to Herb Foster, Marketing Communications Manager for our calculator group. With Herb's agreement, Texas Instruments grants you permission to copy the materials you describe below for the limited purposes you describe below.
Regards,
Larry Bassuk
Copyright Counsel
972-917-5458

-----Original Message-----
From: Rusen Meylani [mailto:meylani@superonline.com]
Sent: Wednesday, February 20, 2002 5:57 PM
To: copyrightcounsel@list.ti.com - Copyright Legal Counsel
Subject: USAGE OF TI 83+ FACILITIES IN MY SAT II MATH BOOKS
Dear Sir,
I am an educational consultant in Istanbul Turkey and I am working with Turkish students who would like to go to the USA for college education. I am writing SAT II Mathematics books where I make use of TI 83+ facilities, screen shots, etc. heavily. Will you please indicate the copyright issues that I will need while publishing my book?
Thanks very much in advance. I am looking forward to hearing from you soon.
Rusen Meylani.

ACKNOWLEDGMENTS

I would like to thank the students from USA, Turkey and other countries for their helpful comments; I have made sure that in this edition we have observed them all.

I would like to thank Mustafa Atakan ARIBURNU, my ex-student and my lifetime friend for believing in me and for his continuous support.

I would like to thank TEXAS INSTRUMENTS for providing scientists and mathematicians such powerful hand - held computers, the TI family of graphing calculators. With these wonderful machines, teachers of mathematics can go beyond horizons without the need to reinvent the wheel all the time. I would also like to thank TEXAS INSTRUMENTS for providing me with a limited copyright to use the graphs that have been produced by the TI 83 / 84 Family of graphing calculators throughout this book.

I would like to thank Erdel VIDORI of TEXAS INSTRUMENTS for his suggestions on the organization and title of this book as well as his invaluable efforts in establishing the link between myself and TEXAS INSTRUMENTS.

I would like to thank Zeynel Abidin ERDEM, the chairman of the Turkish – American Businessmen Association (TABA) for his valuable support and contributions on our projects.

I would like to thank Emel UYSAL and Seda EREN for their valuable contributions.

I would like to thank Pınar ERKORKMAZ for her excellent work on the cover design.

I would like to thank Yorgo İSTEFANOPULOS, Ayşın ERTÜZÜN, Aytül ERÇİL, Bayram SEVGEN, Zeki ÖZDEMİR, and Nuran TUNCALI for whatever I know of analytical thinking.

I would like to thank my mother and my father for being who I am.

I would like to thank my mother in law and my father in law for their continuous support.

Last but not the least, I would like to express my most sincere gratefulness to my wife who has continuously encouraged and supported me throughout every stage of preparing this book. Therefore I dedicate this book to her.

To My Beloved Wife Ebru…

"Seni Seviyorum!"

PREFACE

If you give yourself to mathematics, mathematics will give you the world.

The tests in this book are slightly harder than the actual tests as they are produced to prepare you better for the real test; they **are not realistic in terms of the level of difficulty of the questions**. Please note that if your intention is to get prepared for the real test, this book is the one; but if you are in search of a book that will make you feel good only, you should look elsewhere.

However, we still claim that **the tests in this book as well as the bonus tests in the CD are realistic in the sense that they cover absolutely every type of question which has ever appeared or is likely to appear in the real test**. Understanding how to solve the 900 questions in this book is what you need for a perfect score or for a significant increase in your score in a short period of time.

We have totally **18 tests in this edition that are in accordance with the latest trends in the SAT Math Level 1 Subject Test**. You should also notice that each question is unique; even the ones that might look similar are different versions of the same question type, deliberately placed in the book for the best prep.

A smart test taker who wishes to use this book **should go over each and every question and its solution** without worrying about the time she or he spends. If a student can get a raw score of 40 or more in a test, then he/she is very close to the perfect score in the actual test. One can comfortably add 100 to her/his scaled score in order to estimate what she/he will get in the real test.

Moreover, we have made sure in this edition that the errors or ambiguities related to the use of the English language are eliminated and that each question clearly explains what is given and what is to be found. However we apologize for any errors or vague expressions that might still exist. Please do note that this is a math book and not an English grammar book.

Consequently, **this very book is prepared by experts of SAT Mathematics who intend to help college bound SAT takers in improving their scores in the quickest and the best way possible**.

So, best of luck and be prepared…

Ruşen MEYLANİ

TABLE OF CONTENTS

CHAPTER 1

WHAT SAT MATH SUBJECT TESTS ARE ALL ABOUT

Blank Page

CHAPTER 1 – WHAT SAT MATH SUBJECT TESTS ARE ALL ABOUT

Mathematics Level 1 and Mathematics Level 2 are the two subject tests that the College Board offers. Both tests require at least a scientific, preferably a graphing, calculator. Each test is one hour long. These subject tests were formerly known as the Math Level IC and Math Level IIC subject tests.

Mathematics Level 1 Subject Test

Structure

A Mathematics Level 1 test is made of 50 multiple choice questions from the following topics:

- Algebra and algebraic functions
- Geometry (plane Euclidean, coordinate and three-dimensional)
- Elementary statistics and probability, data interpretation, counting problems, including measures of mean, median and mode (central tendency.)
- Miscellaneous questions of logic, elementary number theory, arithmetic and geometric sequences.

Calculators in the Test

Approximately 60 percent of the questions in the test should be solved without the use of the calculator. For the remaining 40 percent, the calculator will be useful if not necessary.

Mathematics Level 2 Subject Test

Structure

A Mathematics Level 2 test also is made of 50 multiple choice questions. The topics included are as follows:

- Algebra
- Geometry (coordinate geometry and three-dimensional geometry)
- Trigonometry
- Functions
- Statistics, probability, permutations, and combinations
- Miscellaneous questions of logic and proof, elementary number theory, limits and sequences

Calculators in the Test

In Math Level 2, 40 percent of the questions should be solved the without the use of the calculator. In the remaining 60 percent, the calculator will be useful if not necessary.

Which calculator is allowed and which is not:

The simplest reference to this question is this: No device with a QWERTY keyboard is allowed. Besides that any hand held organizers, mini or pocket computers, laptops, pen input devices or writing pads, devices making sounds (Such as "talking" computers) and devices requiring electricity from an outlet will not be allowed. It would be the wisest to stick with TI 84 or TI 89.

Both of these calculators are easy to use and are the choices of millions of students around the world who take SAT exams and also university students in their math courses. It is very important to be familiar with the calculator that you're going to use in the test. You will lose valuable time if you try to figure it out during the test time.

Be sure to learn to solve each and every question in this book. They are carefully chosen to give you handiness and speed with your calculator. You will probably gain an extra 150 to 200 points in a very short period of time.

IMPORTANT: Always take the exam with fresh batteries. Bring fresh batteries and a backup calculator to the test center. You may not share calculators. You certainly will not be provided with a backup calculator or batteries. No one can or will assist you in the case of a calculator malfunction. In such case, you have the option of notifying the supervisor to cancel your scores for that test. Therefore, always be prepared for the worst case scenario (Don't forget Murphy's Rules.)

Number of questions per topics covered
The following chart shows the approximate number of questions per topic for both tests.

Topics	Approximate Number of Questions	
	Level 1	Level 2
Algebra	15	9
Plane Euclidean Geometry	10	0
Coordinate Geometry	6	6
Three-dimensional Geometry	3	4
Trigonometry	4	10
Functions	6	12
Statistics	3	3
Miscellaneous	3	6

Similarities and Differences
Some topics are covered in both tests, such as elementary algebra, three-dimensional geometry, coordinate geometry, statistics and basic trigonometry. But the tests differ greatly in the following areas.

Differences between the tests
Although some questions may be appropriate for both tests, the emphasis for Level 2 is on more advanced content. The tests differ significantly in the following areas:

Geometry

Euclidian geometry makes up the significant portion of the geometry questions in the Math Level 1 test. Though in Level 2, questions are of the topics of coordinate geometry, transformations, and three-dimensional geometry and there are no direct questions of Euclidian geometry.

Trigonometry

The trigonometry questions on Level 1 are primarily limited to right triangle trigonometry and the fundamental relationships among the trigonometric ratios. Level 2 places more emphasis on the properties and graphs of the trigonometric functions, the inverse trigonometric functions, trigonometric equations and identities, and the laws of sines and cosines. The trigonometry questions in Level 2 exam are primarily on graphs and properties of the trigonometric functions, trigonometric equations, trigonometric identities, the inverse trigonometric functions, laws of sines and cosines. On the other hand, the trigonometry in Level 1 is limited to basic trigonometric ratios and right triangle trigonometry.

Functions

Functions in Level 1 are mostly algebraic, while there are more advanced functions (exponential and logarithmic) in Level 2.

Statistics

Probability, mean median, mode counting, and data interpretation are included in both exams. In addition, Level 2 requires permutations, combinations, and standard deviation.

In all SAT Math exams, you must choose the best answer which is not necessarily the exact answer. The decision of whether or not to use a calculator on a particular question is your choice. In some questions the use of a calculator is necessary and in some it is redundant or time consuming. Generally, the angle mode in Level 1 is degree. Be sure to set your calculator in degree mode by pressing "Mode" and then selecting "Degree." However, in Level 2 you must decide when to use the "Degree" mode or the "Radian" mode. There are figures in some questions intended to provide useful information for solving the question. They are accurate unless the question states that the figure is not drawn to scale. In other words, figures are correct unless otherwise specified. All figures lie in a plane unless otherwise indicated. The figures must NOT be assumed to be three-dimensional unless they are indicated to be. The domain of any function is assumed to be set of all real numbers x for which f(x) is a real number, unless otherwise specified.

Important Notice on the Scores

In Level 1 questions the topics covered are relatively less than those covered in the Level 2 test. However, the questions in the Level 1 exam are more tricky compared to the ones in Level 2. This is why if students want to score 800 in the Level 1 test, they have to answer all the 50 questions correctly. But in the Level 2 test, 43 correct answers (the rest must be omitted) are sufficient to get the full score of 800.

Scaled Score	Raw Score in Level 1 Test	Raw Score in Level 2 Test
800	50	43
750	45	38
700	38	33
650	33	28
600	29	22
550	24	16
500	19	10
450	13	3
400	7	0
350	1	-3

CHAPTER 2

MOST ESSENTIAL GRAPHING

CALCULATOR TECHNIQUES

COURTESY   TEXAS INSTRUMENTS

Blank Page

COURTESY ⚹ TEXAS INSTRUMENTS

In this section you will learn the most common graphing calculator techniques along with some critical examples that will raise your scores by at least 50 points. However, there are a lot more to what can be done with the graphing calculator during the test for a typical raise in the score by 150 to 200 points in a very short period of time. Here is a partial snapshot of what you can use your graphing calculator for:

- Polynomial Equations
- Algebraic Equations
- Absolute Value Equations
- Exponential and Logarithmic Equations
- System of Linear Equations, Matrices and Determinants
- Trigonometric Equations
- Inverse Trigonometric Equations
- Polynomial, Algebraic and Absolute Value Inequalities
- Trigonometric Inequalities
- Maxima and Minima
- Domains and Ranges
- Evenness And Oddness

- Graphs of Trigonometric Functions
  - o Period
  - o Frequency
  - o Amplitude
  - o Offset
  - o Axis of wave equation
- Miscellaneous Graphs
- The Greatest Integer Function
- Parametric Graphing
- Polar Graphing
- Limits and Continuity
- Horizontal and Vertical Asymptotes
- Complex Numbers
- Permutations and Combinations

For learning the topics listed above please refer to the following book which is one of a kind:

SAT Math Subject Test with the TI 83 – 84 Family

What is special about the method in this book is that it shortens

40 hours of college preparatory precalculus study **to** an easy **4 hours.**

P.S. We highly recommend the usage of a TI 84 or a TI 89 graphing calculator.

COURTESY   **TEXAS INSTRUMENTS**

Solving Equations

When solving a polynomial or algebraic equation in the form **f(x)=g(x)**, perform the following steps:

i.   Write the equation in the form: **f(x)-g(x)=0**.

ii. Plot the graph of **y=f(x)-g(x)**.

iii. Find the x-intercepts using the **Calc Zero** of TI-84 Plus. However when the graph seems to be tangent to the x-axis at a certain point, you may use the **Calc Min** or **Calc Max** facilities but you should make sure that the y-coordinate of the minimum or maximum point is zero.

Solving Inequalities

When solving an inequality in the form **f(x)<g(x)**, or **f(x)≤g(x)**, or **f(x)>g(x)**, or **f(x)≥g(x)** perform the following steps:

i.   Write the inequality in the form: **f(x)-g(x)<0 or f(x)-g(x)≤0 or f(x)-g(x) >0 or f(x)-g(x)≥0**.

ii. Plot the graph of **y=f(x)-g(x)**.

iii. Find the x-intercepts using the **Calc Zero** of TI-84 Plus. However when the graph seems to be tangent to the x-axis at a certain point, you may use the **Calc Min** or **Calc Max** facilities but you should make sure that the y-coordinate of the minimum or maximum point is zero.

iv. Any value like **-6.61E -10** or **7.2E -11** can be interpreted as 0 as they mean **-6.6x10^{-10}** and **7.2x10^{-11}** respectively.

v.   The solution of the inequality will be the set of values of x for which the graph of f(x)-g(x) lies below the x axis if the inequality is in one of the forms **f(x)-g(x)<0** or **f(x)-g(x)≤0**. The solution of the inequality will be the set of values of x for which the graph of f(x)-g(x) lies above the x axis if the inequality is in one of the forms **f(x)-g(x)>0** or **f(x)-g(x)≥0**. If ≤ or ≥ symbols are involved, then the x-intercepts are also in the solution set.

vi. Please note that the x-values that correspond to asymptotes are never included in the solution set.

Example 1:

$P(x)= 2x^2+3x+1$

$P(a)= 7 \Rightarrow a=?$

Solution:

$2a^2+3a+1=7$

$2a^2+3a-6=0$

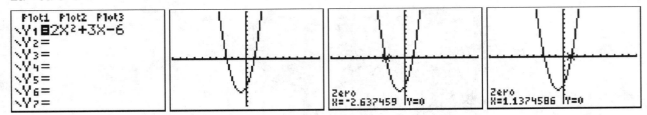

Answer: -2.637 or 1.137

COURTESY 🔷 **TEXAS INSTRUMENTS**

Example 2:

$f(x) = \sqrt{3x+4}$ and $g(x) = x^3$; If is given what $(fog)(x) = (gof)(x)$, then what is x?

Solution:

$(fog)(x) = \sqrt{3x^3 + 4}$

$(gof)(x) = (\sqrt{3x+4})^3$

$\sqrt{3x^3 + 4} = (\sqrt{3x+4})^3$

$\sqrt{3x^3 + 4} - (\sqrt{3x+4})^3 = 0$

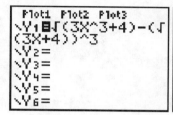 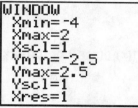 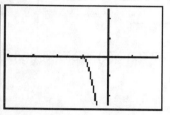

Answer: -1

Example 3:

$|x-3| + |2x+1| = 6 \Rightarrow x = ?$

Solution:

$|x-3| + |2x+1| - 6 = 0$

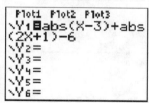 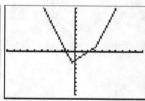 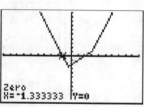 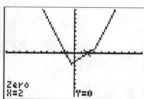

Answer: x = -1.33 or 2

Example 4:

$2^{x+3} = 3^x \Rightarrow x = ?$

Solution:

$2^{x+3} - 3^x = 0$

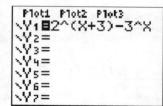 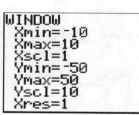 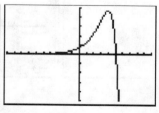 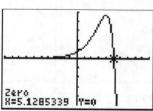

Answer: 5.129

COURTESY ✦ **TEXAS INSTRUMENTS**

Example 5:

$3.281^x = 4.789^y \Rightarrow \dfrac{x}{y} = ?$

Solution:

If $y = 1$ then $\dfrac{x}{y} = x \Rightarrow 3.281^{\,x} = 4.789^{\,1} = 4.789$

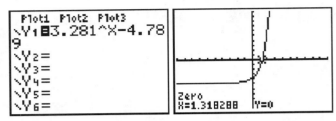

Answer: 1.318

Example 6:

$\log_x 3 = \log_4 x \Rightarrow$ What is the sum of the roots of this equation?

Solution:

$\log_x 3 - \log_4 x = 0$

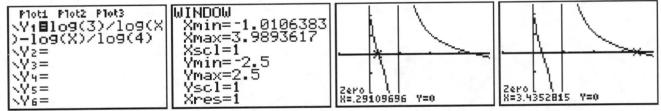

Answer: 0.291+3.435=3.726

Example 7:

$\cos(2x) = 2\sin(90° - x)$. What are all possible values of x between 0° & 360°?

Solution:

Mode: Degrees

$\cos(2x) - 2\sin(90° - x) = 0$

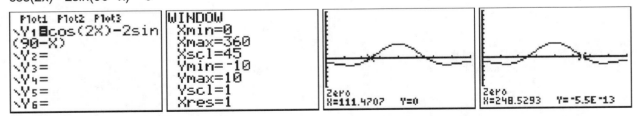

Answer: 111.47°, 248.53°

COURTESY ⬥ TEXAS INSTRUMENTS

Example 8:

2sinx+cos(2x)=2sin²x-1 and $0 \leq x < 2\pi \Rightarrow x=?$

Solution:

Mode: Radians

2sinx+cos(2x) - 2sin²x+1 = 0

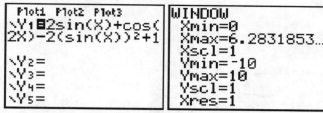

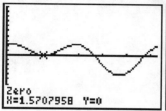 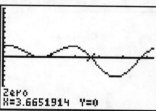 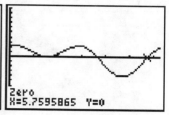

Answer: 1.57, 3.67, 5.76

Example 9:

Solve for x: $\cos^{-1}(2x - 2x^2) = \dfrac{2\pi}{3}$

Solution:

Mode: Radians

$$\cos^{-1}(2x - 2x^2) - \dfrac{2\pi}{3} = 0$$

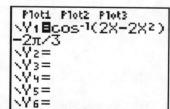 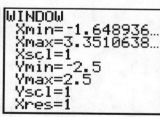 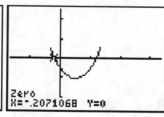 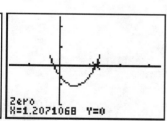

Answer: -0.207 or 1.207

COURTESY **TEXAS INSTRUMENTS**

Example 10:

Solve for x: $\dfrac{|x-2|}{x} > 3$

Solution:

$\dfrac{|x-2|}{x} - 3 > 0$

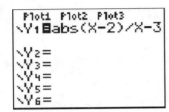 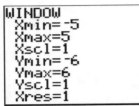 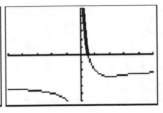

Answer: (0, 0.5)

Example 11:

$|x-2| \le 1$

Solution:

$|x-2| - 1 \le 0$

  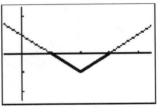

Answer: [1, 3]

Example 12:

$x^2(x-2)(x+1) \ge 0$

Solution:

 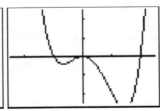 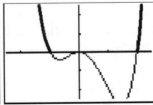

Answer: (-∞, -1] or {0} or [2, ∞)

COURTESY **TEXAS INSTRUMENTS**

<u>**Example 13:**</u>

$\sin(2x) > \sin x$

Find the set of values of x that satisfy the above inequality in the interval $0 < x < 2\pi$.

Solution:

$\sin(2x) - \sin x > 0$

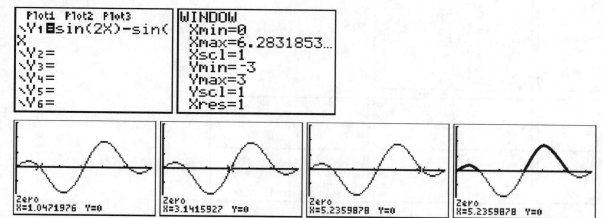

Answer: (0, 1.05) or (3.14, 5.24)

<u>**Example 14:**</u>

$\cos(2x) \geq \cos x$

Find the set of values of x that satisfy the above inequality in the interval $0 \leq x \leq 360°$.

Solution:

$\cos(2x) - \cos x \geq 0$

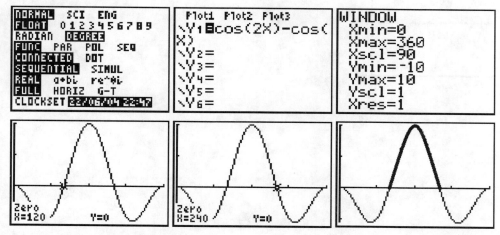

Answer: [120°, 240°] or {0°,360°}

COURTESY **TEXAS INSTRUMENTS**

Miscellaneous Graphing Questions

Example 15:

What is the amplitude, period, frequency, axis of wave and offset of $y=5\sin(x)+12\cos(x)-2$?

Solution:

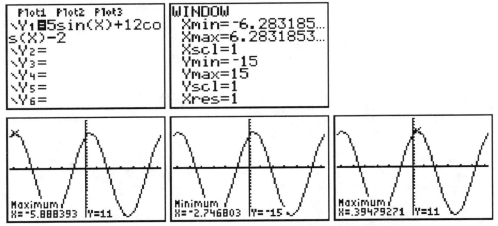

Answer: Amplitude = (11+15) / 2 = 13; Offset: (11 −15) / 2 = -2; Axis of wave: y = -2

Period = 0.394-(-2.747) = 3.141 = π; Frequency = $1/\pi$

Example 16:

Find range of $y=8-2x-x^2$

Solution:

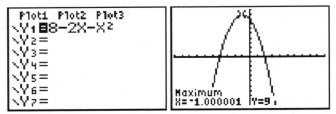

Answer: $y \leq 9$

Example 17:

Find domain and range of the function given by $y= x^{-4/3}$.

Solution:

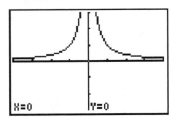

Answer: Domain: $x \neq 0$; Range: $y > 0$.

COURTESY **TEXAS INSTRUMENTS**

Example 18:

Find domain and range of $y=\sqrt{x^2-9}$.

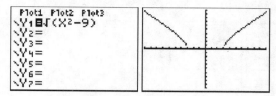

Answer: Domain: $x \leq -3$ or $x \geq 3$; range: $y \geq 0$.

Example 19:

What happens to sinx as x increases from $-\dfrac{\pi}{4}$ to $\dfrac{3\pi}{4}$?

Solution:

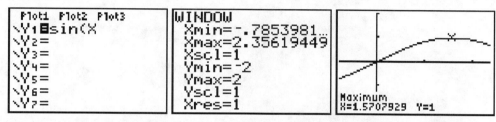

Answer: The function increases between $\pi / 4$ and $\pi / 2$ and then decreases between $\pi / 2$ and $3\pi / 4$.

Example 20:

Find the point of intersection of the graphs y=logx and $y=\ln\dfrac{x}{2}$

Solution:

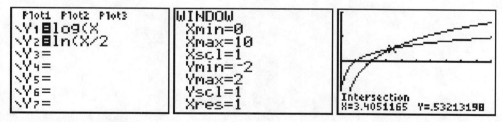

Answer: (3.41, 0.53)

Example 21:

f(x)=2x²+12x+3. If the graph of f(x-k) is symmetric about the y axis, what is k?

Solution:

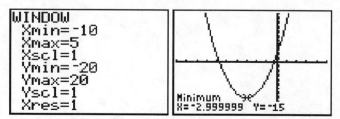

Answer: The graph must be shifted 3 units toward right therefore k=3.

COURTESY ❖ TEXAS INSTRUMENTS

Example 22:

f(x)= -(x-1)^2+3 and -2 ≤ x ≤ 2. Find the range of f(x).

Solution:

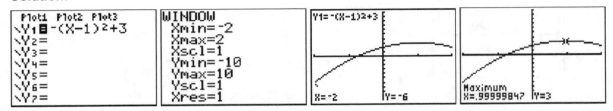

Answer: -6 ≤ y ≤ 3.

COURTESY   ⚜ TEXAS INSTRUMENTS

CHAPTER 3

MODEL TESTS

Blank Page

Model Test 1

Test Duration: 60 Minutes

Directions: For each of the following problems, decide which is the **best** of the choices given. If the exact numerical value is not one of the choices, select the choice that best approximates this value. Then fill in the corresponding oval on the answer sheet.

Notes:

- A calculator will be necessary for answering some (but not all) of the questions in this test. For each question you will have to decide whether or not you should use a calculator. The calculator you use must be at least a scientific calculator; programmable calculators and calculators that can display graphs are permitted.

- The only angle measure used on this test is degree measure. Make sure your calculator is in the degree mode.

- Figures that accompany problems in this test are intended to provide information useful in solving the problems. They are drawn as accurately as possible **except** when it is stated in a specific problem that its figure is not drawn to scale.

- All figures lie in a plane unless otherwise indicated.

- Unless otherwise specified, the domain of any function f is assumed to be the set of all real numbers **x** for which **f(x)** is a real number.

Reference Information: The following information is for your reference in answering some of the questions in this test.

- Volume of a right circular cone with radius **r** and height **h**: $V = \frac{1}{3}\pi r^2 h$

- Lateral area of a right circular cone with circumference of the base **c** and slant height **l**: $S = \frac{1}{2}cl$

- Volume of a sphere with radius **r**: $V = \frac{4}{3}\pi r^3$

- Surface area of sphere with radius **r**: $S = 4\pi r^2$

- Volume of a pyramid with base area **B** and height **h**: $V = \frac{1}{3}Bh$

1. $\dfrac{x}{5} + \dfrac{4x}{5} = 2 \Rightarrow x = ?$

(A) 0.2   (B) 0.8   (C) 2   (D) 2.5   (E) 10

2. $f(x) = x - 3x^3 \Rightarrow f(-3) = ?$

(A) -30   (B) -78   (C) -81   (D) 78   (E) 84

GO ON TO THE NEXT PAGE ▶▶▶

Model Test 1

3. If $7x = 8y$ and $9y = 5z$, then the ratio of x to z is

(A) 5:7   (B) 9:8   (C) 40:63   (D) 56:45   (E) 45:56

4. Which of the following is correct?

(A) $7 + x = 7x$ (B) $3xy - y = 3x$ (C) $11x - x = 11$

(D) $3 \cdot 2 + x = 6 + 3x$ (E) $-3x + y = -(3x - y)$

5. $2x + 2y = -6$ and $3y + 3z = 6$; $z - x = ?$

(A) -5   (B) -3   (C) -2   (D) 2   (E) 5

6. If m and n are natural numbers, which of the following are always natural numbers?

    (I) $m - n$

    (II) $m + n$

    (III) mn

(A) III only   (B) I and II only   (C) II and III only   (D) I and III only   (E) I, II and III

7. For all real values of x, $(5^x)(25^{2x}) = ?$

(A) 5^{5x} (B) 25^{3x} (C) 25^{5x} (D) 125^{2x} (E) 125^{3x}

GO ON TO THE NEXT PAGE ▶▶▶

Model Test 1

8. If the square of the cube root of p is 7, then p can be

(A) – 18.52   (B) – 2.65   (C) 3.66   (D) 49   (E) 343

9. What is the smaller angle between the hands of a clock at 2:20?

(A) 40°   (B) 50°   (C) 60°   (D) 65°   (E) 70°

10. The average of six numbers is 8 and five of the numbers are the first five positive odd integers. What is the remaining number?

(A) 23   (E) 32   (D) 44   (B) 47   (C) 48

11. Figure 1 shows four adjacent squares each having a side of length e making up a bigger square and portion of a circle with point R as center. What is the shaded area in terms of e?

(A) $\dfrac{e^2(4-\pi)}{2}$ (B) $\dfrac{e^2(\pi-4)}{2}$ (C) $\dfrac{e^2(\pi-2)}{2}$ (D) $\dfrac{e^2(\pi-2)}{4}$ (E) $\dfrac{e^2(2\pi-4)}{3}$

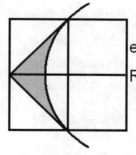

Figure 1

12. A circle passes through the vertices of the regular octagon ABCDEFGH; what is the degree measure of the minor arc AD on this circle?

(A) 45°   (B) 90°   (C) 135°   (D) 180°   (E) 225°

GO ON TO THE NEXT PAGE ▶▶▶

Model Test 1

13. Which of the following geometric figures has exactly four lines of symmetry?

(A) rectangle   (B) circle   (C) square   (D) trapezoid   (E) kite

14. Set X is given by X = {1, 2, 3,...,13} and each number in set X will be placed in a different square among the 13 squares given in figure 2 so that the sum of all numbers in the squares from top left to bottom right is to be the same as the sum of the numbers in the squares from bottom left to top right. Which of the following numbers can be placed in the shaded square?

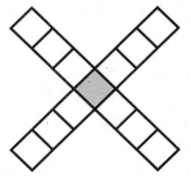

Figure 2

   I. 1   II. 7   III. 13

(A) I only   (B) II only   (C) III only

(D) I and III only   (E) I, II and III

15. What is the domain of the function given by $f(x) = 1 + \sqrt{1 + 3x^2}$?

(A) All real numbers

(B) All real numbers greater than 1

(C) All real numbers greater than or equal to 1

(D) All real numbers greater than or equal to $\dfrac{-1}{3}$

(E) All real numbers greater than or equal to $\dfrac{-1}{\sqrt{3}}$

16. What is the intersection point of the lines m and n given by the equations?   m: 2y − 3x = 4

  (A) (-1, 1)   (B) (1, -2)   (C) (0, 1)   (D) (0, 2)   (E) (2, 0)   n: y = − 5x + 2

GO ON TO THE NEXT PAGE ▶▶▶

Model Test 1

17. The functions f(x) and f(-x) are symmetric across

(A) the origin   (B) the x axis   (C) the y axis   (D) the line given by y = x   (E) the line given by y = -x

18. For the right triangle ABC given in figure 3, m∠B = 2·m∠D and m∠AEC=80°. What is m∠ECD ?

(A) 20°

(B) 30°

(C) 45°

(D) 50°

(E) 60°

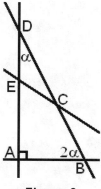

Figure 3
Figure not drawn to scale

19. The seven digit integer given by 3,135,R78 is divisible by 3. R can be

(A) 4   (B) 5   (C) 7   (D) 8   (E) 9

20. If (x, y) are the points on the curve given by $y = x^3 - 5x$, what will be the equation of the reflection of this curve about the y axis?

(A) $y = -x^3 - 5x$ (B) $y = x^3 + 5x$ (C) $y = x^3 - x^2$ (D) $y = -x^3 + 5x$ (E) $y = x^3 - 5x$

21. The cost C in dollars of producing n items is calculated by C = 12.3n + 4.56. What is the maximum number of items that can be produced with $149.95?

(A) 9   (B) 10   (C) 11   (D) 12   (E) 13

GO ON TO THE NEXT PAGE ▶▶▶

Model Test 1

22. What is the area of the shaded region of the semicircle given in figure 4?

(A) 36.37   (B) 72.73   (C) 102.73   (D) 250.47   (E) 500.93

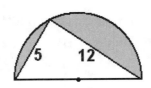

Figure 4
Figure not drawn to scale

23. If in figure 5, sinx = cosy then y can be

(A) 3 only   (B) 1 and 2 only   (C) 2 and 3 only

(D) 4 only   (E) 1 and 3 only

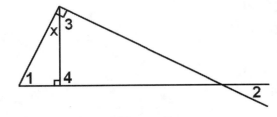

Figure 5
Figure not drawn to scale

24. If the rectangle ABCD given in figure 6 is rotated about point D in the counter clockwise direction for 90° what will be the new coordinates of point B?

(A) (-1, 5)   (B) (-1, 7)   (C) (-2, 5)   (D) (4, -5)   (E) (4, -7)

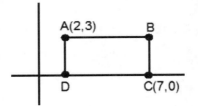

Figure 6

25. If $f(x) = \dfrac{1}{\sin(x)}$ then what is the value of $f(\hat{C})$ if AB = 2BC = 8 and $\hat{B}$ is a right angle for the triangle given in figure 7?

(A) $\dfrac{1}{\sqrt{5}}$ (B) $\dfrac{2}{\sqrt{5}}$ (C) $\sqrt{5}$ (D) $\dfrac{\sqrt{5}}{2}$ (E) 2

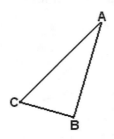

Figure 7

GO ON TO THE NEXT PAGE ▶▶▶

Model Test 1

26. What is the domain of the function given by $f(x) = \dfrac{7x+8}{x^2-9}$?

(A) All real numbers except 3

(B) All real numbers except 3 and -3

(C) All real numbers except $\dfrac{8}{7}$

(D) All real numbers except $-\dfrac{8}{7}$

(E) All real numbers

27. If $f(x) = x^2 - 1$ and $g(x) = \sqrt{x} + 4$, what is the approximate value of $g[f(1.5)]$?

(A) 5.12   (B) 5.50   (C) 6.72   (D) 18.96   (E) 24.72

28. For two integers x and y, A and B are defined by A = x + y and B = x · y; which of the following can be correct?

     I. If A is positive then B can be positive, zero or negative.

     II. If A is negative then B can be positive, zero or negative.

     III. If A is zero then B can be positive, zero or negative.

(A) I only   (B) II only   (C) III only   (D) I and II only   (E) I, II and III

29. If the quadrilateral KLMN given in figure 8 is a parallelogram what are the coordinates of the point N?

(A) (2,5)   (B) (3,6)   (C) (5,6)   (D) (6,1)

(E) It cannot be determined from the given information.

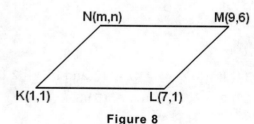

Figure 8

Figure not drawn to scale

GO ON TO THE NEXT PAGE ▶▶▶

Model Test 1

30. It is given that S_1 and S_2 are the lateral surface areas and V_1 and V_2 are the volumes of two similar solids 1 and 2 respectively. If $S_1:S_2 = 4:5$ then what is $V_1:V_2$?

(A) $\dfrac{2}{\sqrt{5}}$ (B) $\dfrac{8}{\sqrt{5}}$ (C) $\dfrac{8}{2\sqrt{5}}$ (D) $\dfrac{5\sqrt{5}}{8}$ (E) $\dfrac{8\sqrt{5}}{25}$

31. In the triangle shown in figure 9, if $\sin x = c$, then $\cos x = ?$

(A) $2\sqrt{1-c^2}$ (B) $2\sqrt{c^2-1}$ (C) $\sqrt{1-c^2}$ (D) $\dfrac{\sqrt{1-c^2}}{2}$ (E) $\dfrac{\sqrt{c^2-1}}{2}$

Figure 9

32. How many members of the set $\{-6, -5, -4, -3, -2, -1, 0, 1, 2, 3, 4, 5, 6\}$ satisfy the inequality given by $x^2 \geq 12 + x$?

(A) 4 (B) 5 (C) 6 (D) 7 (E) 8

33. If $(x + 2)$ is a factor of $P(x) = x^3 - 2x^2 + 3bx - 14$ then $b = ?$

(A) -16 (B) 8 (C) 5 (D) -5 (E) -8

34. If $f(3x - 4) = 4x - 10$ and $f(E) = 2$ then $E = ?$

(A) 2 (B) 3 (C) 5 (D) 9 (E) 12

GO ON TO THE NEXT PAGE ▶▶▶

Model Test 1

35. If each side of a parallelogram is reduced by 20%, what happens to its area?

(A) It increases to 64%.   (B) It decreases to 36%.   (C) It decreases by 36%.

(D) It decreases by 44%.   (E) It decreases to 80%.

36. According to the data given in figure 10, what is the area of the shaded triangle?

(A) 127.3   (B) 127.4   (C) 254.7   (D) 245.7

(E) None of the above

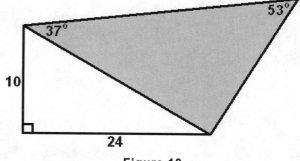

Figure 10
Figure not drawn to scale

37. Neslihan remembers only the first five digits of her seven-digit e-mail password, but she is sure that neither of the last two digits is zero. In order to find out what her password is she has to try out at least how many attempts?

(A) 80   (B) 81   (C) 90   (D) 99   (E) 100

38. Twelve lines in a plane intersect in such a way that each line intersects all others and no intersection point belongs to more than two lines. What is the total number of intersection points?

(A) 12   (B) 66   (C) 72   (D) 132   (E) 144

39. Each interior angle of a regular polygon exceeds the corresponding exterior angle by 100°. What is the number of sides of this polygon?

(A) 6   (B) 8   (C) 9   (D) 12   (E) 15

GO ON TO THE NEXT PAGE ▶▶▶

Model Test 1

40. If the slope of the line passing through (3, p) and (p − 1, 2) is − p then p can be

(A) -0.43   (B) -0.44   (C) -4.25   (D) 2.50   (E) 4.56

41. Which of the following equations, when coupled with the equation 2x − 4y = 3, gives one and only one solution?

(A) x − 2y = 2   (B) 2x + y = 6   (C) 2y = 1 + x   (D) 2x = 4y + 5   (E) 6y − 3x = $-\dfrac{9}{2}$

42. A merchant bought 16 identical leather bags and intended to sell each of the bags at a regular price that is 75% above what he paid for the bag. He sold half of the bags at this price, one fourth at 30% off the regular price and the remaining bags at 60% off the regular price. Which of the following is correct?

(A) His overall loss is 35%.   (B) His overall loss is 36%.

(C) His overall profit is 35%.   (D) His overall profit is 36%.

(E) He collected just the amount he originally paid.

43. Given in figure 11 is a cube having a side of length x and three vertices of the cube labeled as A, B and C. What is the perimeter of triangle ABC in terms of x?

(A) 3x   (B) 3x $\sqrt{2}$ (C) 3x $\sqrt{3}$ (D) 4.2x   (E) 5.2x

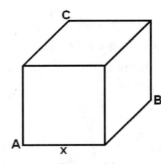

Figure 11

GO ON TO THE NEXT PAGE ▶▶▶

Model Test 1

44. The graph given in figure 12 is the solution set of which of the following inequalities?

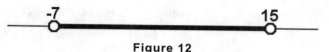

Figure 12

(A) |x − 4| ≤ 11   (B) |x − 4| < 11   (C) |x + 4| < 11

(D) |x + 11| < 4   (E) |x − 11| < 4

45. Which of the following are correct for three coplanar lines l, m, and n?

I. Two distinct lines intersect if they intersect with the same line

II. Two distinct lines are parallel if they are parallel to the same line.

III. Two distinct lines are perpendicular if they are perpendicular to the same line.

(A) I only.   (B) II only.   (C) III only.   (D) II and III only.   (E) I, II and III.

46. Three ladies in a party, Mrs. White, Mrs. Black and Mrs. Gray are wearing white, black and gray hats. Mrs. White says: "Did you notice that none of us is wearing a hat that matches her name?" The lady who is wearing the gray hat says, "Oh, yes, that is correct!" Based on this information which of the following is a correct statement?

(A) Mrs. Gray is wearing the black hat.   (B) Mrs. White is wearing the gray hat.

(C) Mrs. Black is wearing the white hat.   (D) Mrs. White is wearing the black hat.

(E) It is impossible to find who wears which hat.

47. A circle is circumscribed about a triangle with sides of length 5, 12, and 13. What is the radius of this circle?

(A) 1   (B) 6   (C) 6.5   (D) 12   (E) 13

GO ON TO THE NEXT PAGE ▶ ▶ ▶

Model Test 1

48. At most how many spherical rubber balls each having a radius of 1 inch can be placed in a box with the dimensions of 1 foot by 1.5 feet by 2 feet?

(A) 155   (B) 648   (C) 5184   (D) 1237   (E) 1238

49. If log(a) = x and log(b) = y then $\log \dfrac{\sqrt{a}}{b}$ =?

(A) 0.5x + y   (B) 0.5x − y   (C) x + 0.5y   (D) 2x + y   (E) 2x − y

50. Which of the following circles allows the point (3, 2) to be inside it?

(A) $x^2 + (y − 1)^2 = 16$ (B) $(x − 1)^2 + (y − 1)^2 = 4$ (C) $x^2 + y^2 = 13$

(C) $x^2 + (y + 1)^2 = 1$ (D) $(x + 1)^2 + (y + 1)^2 = 25$

S T O P

END OF TEST

(Answers on page 201 − Solutions on page 205)

Model Test 2

Test Duration: 60 Minutes

Directions: For each of the following problems, decide which is the **best** of the choices given. If the exact numerical value is not one of the choices, select the choice that best approximates this value. Then fill in the corresponding oval on the answer sheet.

Notes:

- A calculator will be necessary for answering some (but not all) of the questions in this test. For each question you will have to decide whether or not you should use a calculator. The calculator you use must be at least a scientific calculator; programmable calculators and calculators that can display graphs are permitted.

- The only angle measure used on this test is degree measure. Make sure your calculator is in the degree mode.

- Figures that accompany problems in this test are intended to provide information useful in solving the problems. They are drawn as accurately as possible **except** when it is stated in a specific problem that its figure is not drawn to scale.

- All figures lie in a plane unless otherwise indicated.

- Unless otherwise specified, the domain of any function f is assumed to be the set of all real numbers **x** for which **f(x)** is a real number.

Reference Information: The following information is for your reference in answering some of the questions in this test.

- Volume of a right circular cone with radius **r** and height **h**: $V = \frac{1}{3}\pi r^2 h$

- Lateral area of a right circular cone with circumference of the base **c** and slant height l: $S = \frac{1}{2}cl$

- Volume of a sphere with radius **r**: $V = \frac{4}{3}\pi r^3$

- Surface area of sphere with radius **r**: $S = 4\pi r^2$

- Volume of a pyramid with base area **B** and height **h**: $V = \frac{1}{3}Bh$

1. If $\frac{z+4}{z-3} = \frac{1}{7}$, then z = ?

(A) -5.17 (B) -2.13 (C) 2.13 (D) 3.13 (E) 4.17

2. If $y = 2 \cdot (x - 3)^5$ and x = 5, then y =

(A) 4 (B) 16 (C) 20 (D) 32 (E) 64

GO ON TO THE NEXT PAGE ▶▶▶

Model Test 2

3. A rectangular container is 13 feet high, 22 feet long and, 17 feet deep. How many cubical boxes as shown in figure 1 can be placed in the container?

(A) 18   (B) 180   (C) 14   (D) 140

(E) None of the above

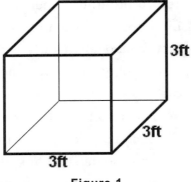

Figure 1

4. If figure 2, if the area of $\triangle ABC$ is 20, then a =

(A) -4   (B) -5   (C) -6   (D) -7   (E) -8

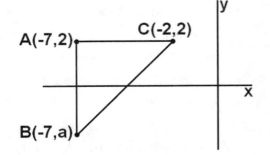

Figure 2

Figure not drawn to scale

5. What is the equation of the line that passes through the point $\left(3, -\frac{1}{2}\right)$ and is parallel to the line

$2y = 3x - 2$?

(A) $2y - 3x = -2$ (B) $2y + 10 = 3x$ (C) $6y = -4x + 9$ (D) $6y - 4x = +9$ (E) $3y - 2x = 10$

6. $2\cdot(2^2 - 1) + 3\cdot(3^2 - 1) + 4\cdot(4^2 - 1) + 5\cdot(5^2 - 1) =$

(A) 14   (B) 80   (C) 94   (D) 210   (E) 238

7. If $x + y = 11$ and $x - y = 7$, then $x^2 - y^2 =$

(A) 4   (B) 7   (C) 15   (D) 49   (E) 77

GO ON TO THE NEXT PAGE ▶▶▶

Model Test 2

8. The block of cubes in figure 3 consists of 60 small cubes. If the block is painted blue, then what percentage of the smaller cubes will have paint on only one of their faces?

(A) 50%   (B) 37%   (C) 18%   (D) 12%   (E) 10%

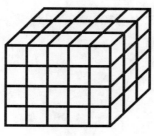

Figure 3

9. If $\sqrt[3]{x^2} = 5$, then $\sqrt{x^3} =$

(A) -5   (B) .2   (C) 11.18   (D) 37.38   (E) 125

10. A wheel with a diameter of length 7 inches makes 12 complete revolutions. How far does the wheel go in inches?

(A) 126   (B) 252   (C) 504   (D) 263.89   (E) 527.79

11. Bige is 10 years younger than Cana who is twice as old as Atakan. If 10 years ago Cana was three times as old as Atakan was, then how old is Bige now?

(A) 10   (B) 20   (C) 30   (D) 40   (E) 50

12. In figure 4, all line segments intersect at right angles. What is the approximate perimeter of the resulting polygon?

(A) 22.15

(B) 25.74

(C) 27.02

(D) 27.29

(E) 29.21

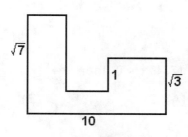

Figure 4

Figure not drawn to scale

GO ON TO THE NEXT PAGE ▶▶▶

Model Test 2

13. If $\sqrt[n]{30^2} = 2^8 \cdot 3^8 \cdot 5^8$, then n = ?

(A) 0.13   (B) 0.25   (C) 0.40   (D) 0.50   (E) 1.00

14. If $P(x) = x^3 + t \cdot x^2 + 5x + 2$ gives the remainder of 8 when divided by x - 3, then what is the value of t?

(A) -6   (B) -4   (C) 0   (D) 2   (E) 8

15. An operation ♦ is defined for all positive real numbers x and y by $x ♦ y = x^y - y^x$. If $2 ♦ a < 0$ then a could equal

   I.   -1

   II. 1

   III. 3

(A) I only   (B) II only   (C) III only   (D) I and III only   (E) I, II and III

16. Gizem had 110 marbles left after giving $\frac{1}{4}$ of her marbles to Mehmet and $\frac{1}{5}$ of her marbles to Nancy. How many marbles did Gizem give to Mehmet?

(A) 35   (B) 40   (C) 50   (D) 55   (E) 90

17. If $x < 0$, then $\left| x + \left| -x \right| + \left| x \right| \right| = ?$

(A) -2x   (B) −x   (C) x   (D) 2x   (E) 3x

GO ON TO THE NEXT PAGE ► ► ►

Model Test 2

18. In figure 5, lines d_1 and d_2 are parallel. If a = 120° and b = 130°, then $\angle\theta$?

(A) 50   (B) 55   (C) 60   (D) 65   (E) 70

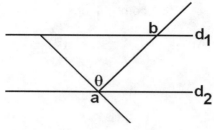

Figure 5
Figure not drawn to scale

19. In figure 6, $\triangle$ ABC is inscribed within a semicircle with center O. If $\angle$ CAB=30° and the circle has diameter 10, then what is the approximate length of arc ADC?

(A) 1.47   (B) 10.4   (C) 10.5   (D) 20.9   (E) 31.4

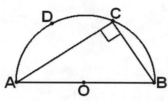

Figure 6
Figure not drawn to scale

20. The fifth root of the square of a number is 4. What is the number?

(A) 8   (B) 16   (C) 32   (D) 64   (E) 128

21. If $f(x) = \dfrac{2}{\sqrt[3]{x+1}}$ for x ≠ -1, then f(7) = ?

(A) .25   (B) 1   (C) 2   (D) 2.83   (E) 5.66

22. What are all values of x for which $|3 - x| \leq 4$?

(A) $x \geq -1$ (B) $x \leq 7$ (C) $-1 \leq x \leq 7$ (D) $x \leq -1$ or $x \geq 7$ (E) x = -1 or x = 7

GO ON TO THE NEXT PAGE ▶ ▶ ▶

Model Test 2

23. A fair coin is flipped 4 times. What is the probability of obtaining exactly 3 tails?

(A) $\dfrac{1}{2}$ (B) $\dfrac{1}{4}$ (C) $\dfrac{1}{6}$ (D) $\dfrac{1}{8}$ (E) $\dfrac{1}{16}$

24. If $i = \sqrt{-1}$, then which of the following must be equal to zero?

(A) $i^{20} + i^{24}$ (B) $i^{14} + 1$ (C) $i^{17} - i^{3}$ (D) $i^{21} + i^{32}$ (E) $i^{15} + i^{28}$

25. Which of the following has the least value?

(A) $\left(\dfrac{1}{2}\right)^{121}$ (B) $\left(\dfrac{1}{4}\right)^{61}$ (C) $\left(\dfrac{1}{8}\right)^{40}$ (D) $\left(\dfrac{1}{16}\right)^{32}$ (E) $\left(\dfrac{1}{32}\right)^{25}$

26. An equation for the circle with its center at the origin and passing through (2, 5) is

(A) $x^2 + y^2 = 29$ (B) $(x - 2)^2 - (y - 5)^2 = 29$ (C) $(x - 2)^2 + (y - 5)^2 = 29$

(D) $x^2 + y^2 = \sqrt{29}$ (E) $(x + 2)^2 + (y + 5)^2 = 29$

27. In figure 7, if $\theta = 4\alpha$, then what is the value of x?

(A) 12.83   (B) 12.31   (C) 9.65   (D) 5.66   (E) 6.15

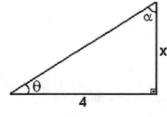

Figure 7
Figure not drawn to scale

GO ON TO THE NEXT PAGE ▶▶▶

28. If a and b are positive odd integers, which one(s) of the following must also be odd?

   I. $a^2 + b^3 + 1$

   II. $2 \cdot (a + b) + 1$

   III. $a + 2 \cdot b + 1$

(A) I only   (B) II only   (C) I and II only

(D) II and III only   (E) I, II and III.

29. $(\sin x + \cos x)^2 + (\sin x - \cos x)^2 = ?$

(A) $4\sin x \cdot \cos x$ (B) $2\sin^2 x$ (C) 2

(D) $4\sin^2 x \cdot \cos^2 x$ (E) $2\cos^2 x$

30. The lengths of 5 pencils are measured and the arithmetic mean is calculated to be 6 inches. The lengths of 3 of these 5 pencils are equal to the median length that is 5 inches. Which of the following must be true?

(A) The longest pencil is 5 inches.

(B) Exactly 1 pencil is longer than the median length.

(C) At least 1 pencil is longer than the median length.

(D) Exactly one pencil is shorter than the median length.

(E) At least one pencil is shorter than the median length.

31. If $g(x) = 2x - 3$ for all x, then the x-intercept of the line given by $y = g\left(\dfrac{x+1}{2}\right)$ is

(A) -3   (B) -2   (C) 0   (D) 1   (E) 2

GO ON TO THE NEXT PAGE ▶▶▶

Model Test 2

32. There are 11 lines on the same plane that intersect in such a way that each line intersects all other lines and each of the resulting points of intersection lies on exactly two distinct lines. If A is the number of all points of intersection and B is the maximum number of intersection points that lie on the same line, then A − B = ?

(A) 44　　　　　(B) 45　　　　　(C) 55　　　　　(D) 100　　　　　(E) 110

33. The surface area of a cube is numerically equal to twice its volume. What is the length of the diagonal of the cube?

(A) 3　　　　　(B) $3\sqrt{2}$　　　　　(C) $3\sqrt{3}$　　　　　(D) $\sqrt{6}$

(E) None of the above

34. Each of the upper and lower bases of a prism is in the shape of a regular n sided polygon. Which of the following are correct statements?

I. Its number of edges is 3n.

II. Its number of vertices is 2n.

III. Its number of faces is n + 2.

(A) I only　　　　(B) II only　　　　(C) III only　　　　(D) II and III only　　　　(E) I, II and III

35. In figure 8, BC // DE. If AE = 3 and EC = 2, then what is the value of $\dfrac{a}{a+b}$?

(A) 3/2　　　　(B) 3/5　　　　(C) 1/2　　　　(D) 3/8　　　　(E) 1/4

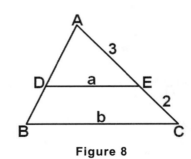

Figure 8

Figure not drawn to scale

GO ON TO THE NEXT PAGE ▶▶▶

Model Test 2

36. If a triangle with vertices A(-1, 5), B(-6, 2), and C(-1, 2) is first reflected across the y-axis and then across the x-axis, which of the following will be the coordinates of the point B?

(A) (-2, 6)　　　　(B) (6, 2)　　　　(C) (6, -2)　　　　(D) (-6, -2)　　　　(E) (2, -6)

37. In the beginning of 1999, the height of a fast growing tropical tree was 10.12 feet. If the height of the tree increases by 115% each year, what will its height be in the beginning of 2004?

(A) 20.35　　　　(B) 23.05　　　　(C) 46.5　　　　(D) 58.19　　　　(E) 464.9

38. If θ is a positive acute angle and $\cos^2\theta = \dfrac{a}{b}$, then $\sin\theta =$

(A) $\sqrt{\dfrac{a}{b}-1}$　　(B) $\sqrt{\dfrac{b}{a}-1}$　　(C) $\sqrt{1-\dfrac{a}{b}}$　　(D) $\sqrt{1-\dfrac{b}{a}}$　　(E) $\sqrt{1+\dfrac{a}{b}}$

39. Figure 9 shows three views of the same cube. If the cube is opened and flattened which of the following can be the resulting pattern?

Figure 9

(A)　　　　　　(B)　　　　　　(C)　　　　　　(D)

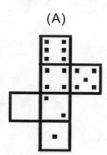　　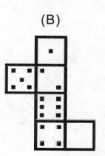　　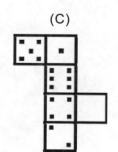　　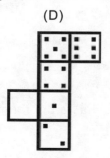

(E) None of the above

GO ON TO THE NEXT PAGE ▶▶▶

Model Test 2

40. What is the domain of the function given by $f(x) = \dfrac{\sqrt{x+2}}{x-1}$?

(A) x > -2 and x ≠ 1   (B) -2 ≤ x < 1 or x > 1   (C) x > -2

(D) -2 ≤ x < 1 and x > 1   (E) -2 < x < 1 and x > 1

41. In figure 10, if a right circular cylinder is inscribed in a cube whose volume is 64, then the volume of the cylinder is approximately

(A) 13.73   (B) 25.13   (C) 36.53   (D) 50.27   (E) 100.53

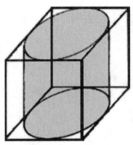

Figure 10

42. In figure 11, the area of the parallelogram ABCD is 160. What is the perimeter of the parallelogram?

(A) 31   (B) 62   (C) 31.31   (D) 62.62   (E) 62.63

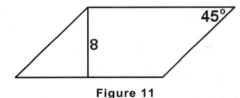

Figure 11

Figure not drawn to scale

43. What is the range of the function f(x) = (x − 2)^2 + 2 defined for 1 ≤ x ≤ 4?

(A) 0 ≤ y ≤ 2   (B) 3 ≤ y ≤ 6   (C) 2 ≤ y ≤ 4   (D) 2 ≤ y ≤ 6   (E) 4 ≤ y ≤ 6

GO ON TO THE NEXT PAGE ▶▶▶

44. The n-th term in the sequence of numbers $\{-2, 4, 10, 16, 22, ...\}$ is

(A) $n - 3$ (B) $6n - 4$ (C) $2(n - 2)$ (D) $3 \cdot (2n - 1) - 5$ (E) $(-1)^n \cdot (8 - 6n)$

45. For the right square pyramid given in figure 12, perimeter of base and altitude are 24 and 8 feet respectively. What is the perimeter of the shaded triangle?

(A) 24.8   (B) 25.2   (C) 26.6   (D) 27.4

(E) It cannot be determined from the information given.

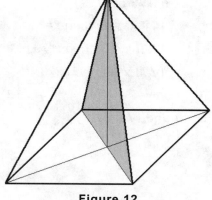

Figure 12

46. For how many distinct values of p does solution set of the equation $(2p - 1)x^2 - 5x + 4 = 0$ contain one and only one real number?

(A) 0   (B) 1   (C) 2   (D) 3   (E) more than 3

47. In figure 13, the radius of the circle is 10 and the length of chord AB is $10\sqrt{3}$; then what is the area of the shaded region?

(A) 17.32   (B) 20.94   (C) 61.42   (D) 78.54   (E) 104.72

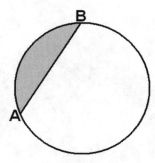

Figure 13

GO ON TO THE NEXT PAGE ▶ ▶ ▶

48. If the line x = a is tangent to the circle defined by $(x - 2)^2 + y^2 = 9$, then a can be

(A) -1 only   (B) 5 only   (C) -1 or 5   (D) 3 only   (E) -3 or 3

49. The statement given by "If x=3, then x^2=9" is logically equivalent to which one(s) of the following?

     I. $x \neq 3$ or x^2=9

     II. if x^2 = 9, then x=3.

     III. If $x \neq 3$, then $x^2 \neq 9$.

     IV. If $x^2 \neq 9$, then $x \neq 3$.

(A) IV only   (B) I and IV only   (C) II and IV only

(D) III and IV only   (E) I, II, III and IV

50. Diameter of the semicircle and one side of square ACDF coincide as shown in figure 14. If BE is perpendicular to DF and 4DE = EF = 16, then what is the length of BE?

(A) 24   (B) 26   (C) 28   (D) 30

(E) It cannot be determined from the given information.

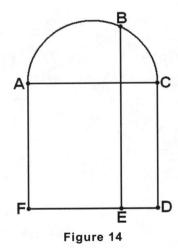

Figure 14
Figure not drawn to scale

S T O P

END OF TEST

(Answers on page 201 – Solutions on page 207)

Model Test 3

Test Duration: 60 Minutes

Directions: For each of the following problems, decide which is the **best** of the choices given. If the exact numerical value is not one of the choices, select the choice that best approximates this value. Then fill in the corresponding oval on the answer sheet.

Notes:

- A calculator will be necessary for answering some (but not all) of the questions in this test. For each question you will have to decide whether or not you should use a calculator. The calculator you use must be at least a scientific calculator; programmable calculators and calculators that can display graphs are permitted.

- The only angle measure used on this test is degree measure. Make sure your calculator is in the degree mode.

- Figures that accompany problems in this test are intended to provide information useful in solving the problems. They are drawn as accurately as possible **except** when it is stated in a specific problem that its figure is not drawn to scale.

- All figures lie in a plane unless otherwise indicated.

- Unless otherwise specified, the domain of any function f is assumed to be the set of all real numbers **x** for which **f(x)** is a real number.

Reference Information: The following information is for your reference in answering some of the questions in this test.

- Volume of a right circular cone with radius **r** and height **h**: $V = \frac{1}{3}\pi r^2 h$

- Lateral area of a right circular cone with circumference of the base **c** and slant height **l**: $S = \frac{1}{2}cl$

- Volume of a sphere with radius **r**: $V = \frac{4}{3}\pi r^3$

- Surface area of sphere with radius **r**: $S = 4\pi r^2$

- Volume of a pyramid with base area **B** and height **h**: $V = \frac{1}{3}Bh$

1. If a - b - 4 = 7, then 4a - 4b=

(A) 4   (B) 11   (C) 22   (D) 44   (E) 77

2. If $f(x) = 5x^2 - 3x + 4$, then $f\left(\frac{2}{5}\right) =$

(A) 1.50   (B) 3.60   (C) 4.12   (D) 6.80   (E) 8.50

GO ON TO THE NEXT PAGE ▶▶▶

Model Test 3

3. If Zehra starts reading a 210 page book at 08:00 AM at the rate of 3 minutes per page, then when can she finish reading the book?

(A) 06:00 PM   (B) 06:30 PM   (C) 06:30 AM   (D) 06:50 PM   (E) 06:50 AM

4. If $5 \cdot 3^{n+3} = 1215$, then n=

(A) -1   (B) 1   (C) 2   (D) 3   (E) 20

5. If the circle given by $(x-1)^2+(y+2)^2=R^2$ is tangent to the line $4x-3y=-10$ then R=?

(A) 4   (B) 5   (C) 7   (D) 10

(E) None of the above

6. When factored $2x^2 - 5x - 3$ equals which of the following?

(A) $(x + 1)(2x - 3)$ (B) $(2x - 1)(x - 3)$ (C) $(x + 1)(2x + 3)$ (D) $(2x + 1)(x - 3)$ (E) $(2x - 1)(x + 3)$

7. If $x < 0$ and $x^2 = 9$ then $x^5 =$

(A) -243   (B) -81   (C) 3   (D) 81   (E) 243

8. If $f(x) = \left| x^2 - 1 \right|$ and $g(x) = \dfrac{1}{x^3}$, then $f(g(-1)) =$

(A) -1   (B) 0   (C) 1   (D) 2   (E) 3

GO ON TO THE NEXT PAGE ▶▶▶

Model Test 3

9. If $x = 4$, then $\dfrac{3 \cdot x^{78}}{5 \cdot x^{76}} =$

(A) 0.62   (B) 1.67   (C) 4.80   (D) 9.60   (E) 38.4

10. If $2x^3 + 3x^2 - 11x - 6 = 0$ and $2x^3 - 11x^2 + 17x - 6 = 0$, then x can be

(A) 0.5   (B) -0.5   (C) 3   (D) -2   (E) 2

11. $(x - y)^2 = 25$ and $(x + y)^2 = 49$, then xy?

(A) 1   (B) 4   (C) 6   (D) 12   (E) 24

12. EB is the bisector of angle AEC and the length of segment BA is 12 inches as shown in figure 1. What is the length in inches of segment FA?

(A) 11   (B) 12   (C) $6\sqrt{3}$ (D) $7\sqrt{2}$

(E) It cannot be determined from the information given.

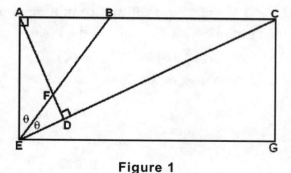

Figure 1

13. If x and y are positive integers, then which of the following conditions must be satisfied so that $x \cdot (y + 1)$ will represent an odd integer?

    I. x^2 is odd

    II. y^2 is even

    III. $x + y + 1$ is odd

(A) I only   (B) II only   (C) III only   (D) I and II only   (E) I and III only

GO ON TO THE NEXT PAGE ▶▶▶

14. If $x = 3\cos\theta$ and $y = \sin\theta$, then which of the following must be correct?

(A) $x^2 + 9y^2 = 1$ (B) $\dfrac{x^2}{9} + y^2 = 3$ (C) $\dfrac{y^2}{9} + x^2 = 1$ (D) $x^2 + y^2 = 9$ (E) $\dfrac{x^2}{9} + y^2 = 1$

15. The fourth root of the cube of a number is 24. Which of the following is the number?

(A) 10.84 (B) 18.40 (C) 36.80 (D) 69.23 (E) 72.14

16. The line $2x - 3y = 11$ is perpendicular to which of the following lines?

(A) $2x - 3y = 7$ (B) $2x + y = 11$ (C) $3x + 2y = 8$ (D) $2x + 3y = 9$ (E) $-3x + 2y = 10$

17. In figure 2, arc EF and central angle ABF measure 80°
and 60° respectively. What is the measure of angle CDE?

(A) 10° (B) 20° (C) 30° (D) 40°

(E) It cannot be determined from the information given.

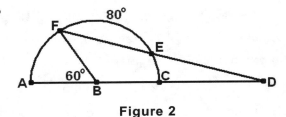

Figure 2

18. What is the remainder when $2x^5 + 4x^3 - x + 1$ is divided by $x - 1$?

(A) -4 (B) 2 (C) 6 (D) 10 (E) 11

GO ON TO THE NEXT PAGE ▶▶▶

Model Test 3

19. If every member of a library borrowed at least one book, then which of the following cannot be the average number of books borrowed per person in one month based on the data given in the table?

(A) 16.15   (B) 15.95   (C) 15.88

(D) 15.81   (E) 15.74

Number of People	Number of Books Borrowed per Month
42	20
26	18
22	15
14	8
8	Less than 6

20. Which of the following is the solution set to the inequality $2x^2 + x - 1 \leq 0$?

(A) $\dfrac{-1}{2} \leq x \leq 1$ (B) $-1 \leq x \leq \dfrac{1}{2}$ (C) $x \leq -1$ or $x \geq \dfrac{1}{2}$ (D) $x \leq -\dfrac{1}{2}$ or $x \geq 1$ (E) All real values of x

21. For a convex pentagon the measures of the interior angles are in the ratio of 2:3:4:5:6. What is the measure of the greatest angle?

(A) 72°   (B) 108°   (C) 144°   (D) 135°   (E) 162°

22. When twice a number n is subtracted from its square, the result is 3. Which of the following can be the number n?

(A) -3   (B) -1   (C) 0   (D) 1   (E) 2

23. If $x^3 = -y^2$ and y is nonzero then which of the following statements must be correct?

　　　I. x is negative   II. x + y is positive   III. x − y is negative

(A) I only   (B) II only   (C) III only   (D) I and II only   (E) I and III only

GO ON TO THE NEXT PAGE ▶▶▶

Model Test 3

24. If $0° \leq x \leq 90°$, and $\sqrt{2} \cdot \sin^2 x = 2 \cdot \sin x - \sqrt{2} \cdot \cos^2 x$, then $\tan x =$

(A) 0   (B) 1   (C) $\sqrt{2}$ (D) 2   (E) $2\sqrt{2}$

25. Sum of the numerical values of the perimeter and the area of a square equals 77. What is the length of one of the sides of the square?

(A) 7   (B) 11   (C) 14   (D) 28   (E) 49

26. If $f(x) = 3x - 2$ and $f(g(3)) = 16$, then which of the following can be $g(x)$?

(A) $x - 3$ (B) $x + 6$ (C) $x^2 - 3$ (D) $2x - 3$ (E) $2x + 3$

27. If $x + y = 9$ and $x^2 - y^2 = 45$, then $\dfrac{2x}{y} =$

(A) 3   (B) 3.5   (C) 5   (D) 7   (E) 10

28. If the exterior angles of a triangle are in ratio 2:3:4. What is the measure of the smallest interior angle?

(A) 20°   (B) 40°   (C) 80°   (D) 100°   (E) 120°

GO ON TO THE NEXT PAGE ▶▶▶

Model Test 3

29. If the origin is the midpoint of the line segment whose end points are given by (2, -3) and (a, b), then a + b =

(A) -5   (B) -1   (C) 0   (D) 1   (E) 5

30. The average height of 7 people is 72 inches. The average height of 6 other people is 69 inches. What is the average height of all 13 people?

(A) 69   (B) 70.92   (C) 70.52   (D) 70.62   (E) 72

31. If the range of the function $f(x) = \dfrac{3x - 1}{5}$ is given by -2≤y≤1, then what is the domain of the function?

(A) -1≤x≤1   (B) -3≤x≤2   (C) -2≤x≤3   (D) -1≤x≤2   (E) -1≤x≤0

32. If $f(x) = \sqrt{x} + 1$ and $g(x) = x^2$, then which of the following could be a portion of the graph of f(g(x))?

(A)	(B)	(C)	(D)	(E)
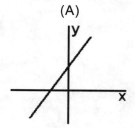	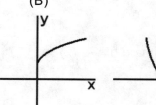	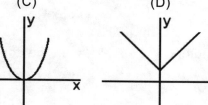		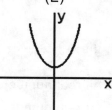

GO ON TO THE NEXT PAGE ▶▶▶

33. If a parallelogram has vertices (-2, 0), (4, 0), (5, 4), and (a, b), then a + b can be

     I. 3   II. -7   III. 15

(A) I only   (B) II only   (C) III only   (D) I and II only   (E) I, II and III

34. A function f(x) has the property that f(-x) = f(x) for all real values of x. If the point (5, -25) is on the graph of f(x), then which of the following points must also be on the same graph?

(A) (-5, 25)   (B) (-5, -25)   (C) (0, 25)   (D) (5, 25)   (E) (25, -5)

35. What is the maximum value of the function $f(x) = -1 + 4x - 2x^2$ over the interval $-1 \leq x \leq 2$?

(A) -7   (B) -1   (C) 1   (D) 2.4   (E) It has no maximum value over the specified interval.

36. f(x) is defined as the remainder when [x] is divided by 7, where the symbol [x] represents the greatest integer less than or equal to x; f(101.5) + f(34.4) − f(133.6) = ?

(A) 3   (B) 5   (C) 6   (D) 8   (E) 9

37. If $f(x) = x^2 + 4x - 5$ and $g(x) = 2 - x - x^2$, then the domain of $\left(\dfrac{f}{g}\right)(x)$ is

(A) All real numbers.

(B) All real numbers -2, 1 and 5.

(C) All real numbers except for 1.

(D) All real numbers except for -2.

(E) All real numbers except for -2 and 1.

GO ON TO THE NEXT PAGE ▶▶▶

Model Test 3

38. The angle of elevation of a tree measured from a point 75 feet away from its base is 27°. What is the height of the tree in feet?

(A) 34.05   (B) 38.21   (C) 66.83   (D) 75   (E) 147.2

39. If a rectangle with sides AB = 5 and BC = 12 is rotated about side BC for 180°, then what will be volume of the resulting object?

(A) 314.16   (B) 628.12   (C) 753.98   (D) 471.24   (E) 942.48

40. An equilateral triangle and a square have equal areas. What is the ratio of the perimeter of the triangle to the perimeter of the square?

(A) 1.14   (B) 1.25   (C) 1.32   (D) 1.41   (E) 1.52

41. A positive integer n greater than 13 is divided by 13 and it is given that the remainder of the division is nonzero. What is the probability that the remainder will be a prime number?

(A) 0   (B) $\dfrac{5}{12}$ (C) $\dfrac{1}{2}$ (D) $\dfrac{1}{12}$ (E) 1

42. The acceleration of a freely falling object in air is directly proportional to the square of its velocity. If the acceleration of an object is 6 m/s^2 when its velocity is 46 m/s, what will its velocity be in m/s when the acceleration is 12 m/s^2?

(A) 92 m/s   (B) 75 m/s   (C) 65 m/s   (D) 60 m/s   (E) 33 m/s

GO ON TO THE NEXT PAGE ▶▶▶

43. If a cube has a diagonal of length d, which of the following represents the surface area of the cube in terms of d?

(A) $\dfrac{\sqrt{3}d^3}{9}$ \qquad (B) $\dfrac{d^3}{9}$ \qquad (C) $\dfrac{d^2}{3}$ \qquad (D) d^2 \qquad (E) $2d^2$

44. $f(x)=3x^2$; $f(\log_7 4)=?$

(A) 0.71 \qquad (B) 1.52 \qquad (C) 1.37 \qquad (D) 1.09

(E) None of the above

45. $\cos\alpha\cdot\left(\tan\alpha+\dfrac{1}{\tan\alpha}\right)=$

(A) $\dfrac{1}{\cos\alpha}$ \qquad (B) $\dfrac{1}{\sin\alpha}$ \qquad (C) 1 \qquad (D) $\sin\alpha$ \qquad (E) $\cos\alpha$

46. In figure 3, a circle with center O is inscribed within square ABCD. If the area of the circle is 16π, which of the following is the area of the shaded region?

(A) $16 - 4\pi$ \quad (B) $16 + 4\pi$ \quad (C) $64 - 4\pi$ \quad (D) $64 + 4\pi$ \quad (E) $64 - 16\pi$

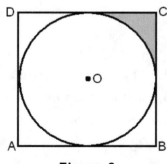

Figure 3

GO ON TO THE NEXT PAGE ▶▶▶

Model Test 3

For questions 47 – 48 please refer to the data given in the following graph that shows the number of books checked by the crew working in **RUSH** Publications.

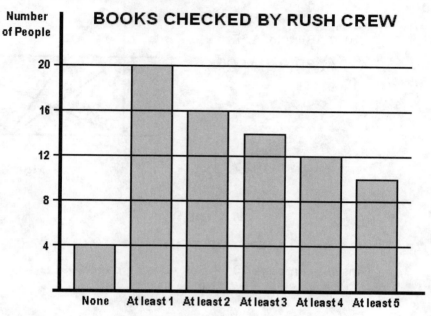

47. How many members of the **RUSH** crew checked exactly 3 books?

(A) 2   (B) 4   (C) 6   (D) 12   (E) 14

48. How many people are there in the **RUSH** crew totally?

(A) 20   (B) 24   (C) 54   (D) 66   (E) 74

49. In figure 4, if $AD \cdot DB = 49$, then what is the value of DC?

(A) 6   (B) 7   (C) 13   (D) 42

(E) There is not enough information to find DC.

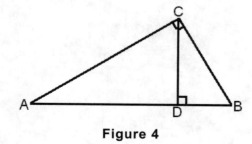

Figure 4

GO ON TO THE NEXT PAGE ▶▶▶

Model Test 3

50. If d_2 is equidistant from the parallel lines d_1 and d_3 in figure 5, which of the following triangles are equal in area?

(A) $\triangle BFD$ and $\triangle FED$

(B) $\triangle ABF$ and $\triangle FED$

(C) $\triangle BFD$ and $\triangle BCD$

(D) $\triangle ABD$ and $\triangle CDE$

(E) $\triangle ACD$ and $\triangle CDE$

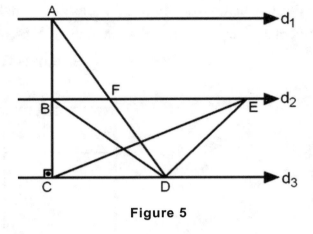

Figure 5

S T O P

END OF TEST

(Answers on page 201 – Solutions on page 209)

Model Test 4

Test Duration: 60 Minutes

Directions: For each of the following problems, decide which is the **best** of the choices given. If the exact numerical value is not one of the choices, select the choice that best approximates this value. Then fill in the corresponding oval on the answer sheet.

Notes:

- A calculator will be necessary for answering some (but not all) of the questions in this test. For each question you will have to decide whether or not you should use a calculator. The calculator you use must be at least a scientific calculator; programmable calculators and calculators that can display graphs are permitted.

- The only angle measure used on this test is degree measure. Make sure your calculator is in the degree mode.

- Figures that accompany problems in this test are intended to provide information useful in solving the problems. They are drawn as accurately as possible **except** when it is stated in a specific problem that its figure is not drawn to scale.

- All figures lie in a plane unless otherwise indicated.

- Unless otherwise specified, the domain of any function f is assumed to be the set of all real numbers **x** for which **f(x)** is a real number.

Reference Information: The following information is for your reference in answering some of the questions in this test.

- Volume of a right circular cone with radius **r** and height **h**: $V = \frac{1}{3}\pi r^2 h$

- Lateral area of a right circular cone with circumference of the base **c** and slant height **l**: $S = \frac{1}{2}cl$

- Volume of a sphere with radius **r**: $V = \frac{4}{3}\pi r^3$

- Surface area of sphere with radius **r**: $S = 4\pi r^2$

- Volume of a pyramid with base area **B** and height **h**: $V = \frac{1}{3}Bh$

1. If $x + 3x + 5x = 2x + 4x + 6x - 12$, then $2x =$

(A) 2   (B) 3   (C) 4   (D) 6   (E) 8

2. For all $x \neq 0$, $\dfrac{\frac{x}{9}}{x} = ?$

(A) $x^2 / 9$ (B) $-x / 3$ (C) $x/3$ (D) $1/9$ (E) $x/9$

GO ON TO THE NEXT PAGE ▶▶▶

Model Test 4

3. If x= -1, then $(1 - x) \cdot (x - 1)$

(A) 4   (B) 2   (C) 1   (D) 0   (E) − 4

4. For the parallelogram ABCD given in figure 1, what is the value of n − m?

(A) 0

(B) 2

(C) 4

(D) 6

(E) 8

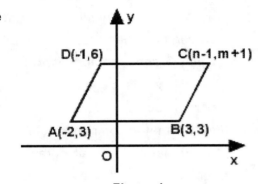

Figure 1

Figure not drawn to scale

5. $(a + b + 2) \cdot (a − b + 2)=$

(A) $a^2 − 2ab + b^2$ (B) $(a + 2)^2 − b^2$ (C) $a^2 − (b+2)^2$ (D) $a^2 + 2ab + b^2$ (E) $(a + b)^2 − 4$

6. At what point does the graph $\dfrac{2}{3}x - \dfrac{3}{4}y = \dfrac{4}{5}$ intersect the x - axis?

(A) -4/3   (B) -3/4   (C) -8/15   (D) 4/5   (E) 6/5

7. If 3xy= 5, then $9x^2y^2=$

(A) 15   (B) 25   (C) 27   (D) 45   (E) 75

GO ON TO THE NEXT PAGE ▶▶▶

Model Test 4

8. If lines d_1 and d_2 are parallel and are intersected by line d_3, what is the sum of the measures of the exterior angles on the same side of line d_3?

(A) 60°　　　　(B) 90°　　　　(C) 120°　　　　(D) 180°　　　　(E) 270°

9. If $x^2 = y^2 + 33$ and $x - y = 3$, then $y=$?

(A) 1　　　　(B) 3　　　　(C) 4　　　　(D) 7　　　　(E) 11

10. If the square root of the cube root of a number is $\frac{1}{2}$ what is the number?

(A) $\frac{1}{256}$　　　　(B) $\frac{1}{64}$　　　　(C) $\frac{1}{36}$　　　　(D) $\frac{1}{32}$　　　　(E) $\frac{1}{16}$

11. How many four digit integers have distinct digits that sum up to 6?

(A) 12　　　　(B) 15　　　　(C) 18　　　　(D) 21　　　　(E) 24

12. For the cubes A, B and C, the volume of A is three times that of B and the volume of B is two thirds that of C. If A has a volume of 216 cubic meters, what is the difference of volumes of B and C, in cubic meters?

(A) 36　　　　(B) 72　　　　(C) 108　　　　(D) 144　　　　(E) 180

GO ON TO THE NEXT PAGE ▶▶▶

13. In figure 2, if line k is rotated clockwise 30 degrees about point A, it will be perpendicular to line m. Before this rotation, the triangle that is formed by the lines k, l and m can be classified as:

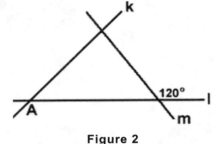

Figure 2
Figure not drawn to scale

    I. Acute

    II. Obtuse

    III. Equilateral

(A) I only   (B) II and III only

(C) III only   (D) I and III only

(E) II only

14. If $3x + 2x + x = y$, then $2x - y =$

(A) 5x   (B) 4x   (C) -4x   (D) -5x   (E) -11x

15. If $f(x) = \dfrac{1}{x^2}$ for $x \neq 0$, then $f\left(\dfrac{1}{2}\right) =$

(A) 0.25   (B) 0.5   (C) 1   (D) 2   (E) 4

16. If $20 \cdot 10^n = 2^5 \cdot 5^4$, what is the value of n?

(A) 3   (B) 4   (C) 5   (D) 9   (E) 10

17. Which of the following values of x satisfies $|x-1| \geq 3$?

(A) − 1   (B) 0   (C) 2   (D) 3   (E) 4

GO ON TO THE NEXT PAGE ▶▶▶

Model Test 4

18. A rectangle is drawn in the Cartesian plane and the vertices of this rectangle are A(1,1), B(6,1), C(6,5) and D(1,5). When a point is selected randomly within the rectangle, what is the probability that the x – coordinate of the selected point is less than its y – coordinate?

(A) 33%   (B) 40%   (C) 50%   (D) 60%   (E) 66%

19. In figure 3, points A, B, and C are equally spaced on the circle centered at P and having the radius of 6. What is the perimeter of the shaded region?

(A) 22.39   (B) 12.57   (C) 22.96   (D) 11.48   (E) 10.39

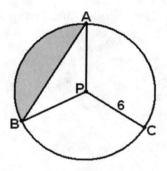

Figure 3

20. If θ is a positive acute angle and $\tan \theta = A$, then $\sin \theta$ is equal to

(A) $\dfrac{1}{A}$ (B) $\dfrac{A}{\sqrt{A^2+1}}$ (C) $-\dfrac{A}{\sqrt{A^2+1}}$ (D) $\dfrac{\sqrt{A^2+1}}{A}$ (E) $-\dfrac{\sqrt{A^2+1}}{A}$

21.   X = {1, 3, 5, 7, 9}

Y = {2, 4, 6, 8}

A = {a, b, c, d, e, f, g, h}

A company uses id codes for its employees. Each code consists of a two – digit number followed by a single letter, of the form x-y-a. Shown above are the sets X, Y, and A from which the digits x and y, and the letter a respectively, are chosen. How many possible id codes are there?

(A) 17   (B) 28   (C) 72   (D) 160   (E) 576

GO ON TO THE NEXT PAGE ▶▶▶

Model Test 4

22. A fair coin is flipped 3 times. What is the probability of obtaining at least 2 heads?

(A) $\frac{1}{2}$ (B) $\frac{1}{4}$ (C) $\frac{3}{8}$ (D) $\frac{5}{8}$

(E) None of the above

23. In figure 4, $\triangle ABC$ is equilateral. If each side of $\triangle ABC$ is 6 inches long, then what is the perimeter of the shaded region?

(A) 4   (B) 6   (C) 9

(D) 12   (E) 18

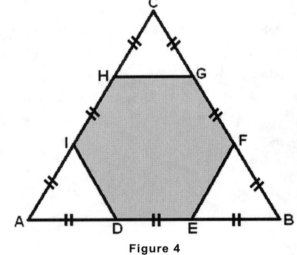

Figure 4

Figure not drawn to scale

24. If $i^2 = -1$ and $i + i^2 + \ldots\ldots + i^k = 0$, then the least positive integer value of k is

(A) 3   (B) 4   (C) 5   (D) 6   (E) 8

25. Given in figure 5 is a circle having its center at O and having the radius of length R. If the measure of θ is 63 degrees and R = 2.5 then what is the length of the major arc AB?

(A) 12.8   (B) 12.9   (C) 13.0   (D) 13.1   (E) 13.2

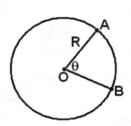

Figure 5

GO ON TO THE NEXT PAGE ▶▶▶

Model Test 4

26. The total length of m rods each m inches long is A and the total length of n rods each n inches long is B. If A and B differ by m + n, m and n can be respectively

(A) 5 and 7   (B) 5 and 8   (C) 8 and 9   (D) 9 and 12   (E) 10 and 13

27. Video store A charges $40 for membership and the rental cost per DVD is $5. Video store B charges $7 per DVD and requires no payment for membership. Video store A is more profitable for a customer who rents

(A) less than 10 movies all at once.   (B) less than 15 movies all at once.

(C) less than 20 movies all at once.   (D) exactly 20 movies all at once.

(E) more than 20 movies all at once.

28. Which of the following is equivalent to x = y?

    I. $x^3 = y^3$

    II. $|x| = |y|$

    III. $\sqrt{x} = \sqrt{y}$

(A) I only   (B) I and II only   (C) I and III only   (D) II and III only   (E) I, II and III

29. The depth of a circular pool in yards is a function of the horizontal distance r yards from the center of the pool given by $d(r) = \dfrac{1}{6}(r^2 - 4r + 7)$. Which of the following is the closest approximation of the depth of the pool, in feet, at a horizontal distance of 2 yards from the center of the pool?

(A) 1/2   (B) 1   (C) 3/2   (D) 2   (E) 5/2

GO ON TO THE NEXT PAGE ▶ ▶ ▶

Model Test 4

30. In the xy plane, the points O(-1,0), P(5,2), R(-4,-2) and S(3,-4) can be connected to form line segments. Which of the following pair of segments have the same lengths?

(A) OP and PR   (B) OP and PS   (C) OP and RS   (D) PR and RS   (E) PR and PS

31. Line m has a negative slope and a positive y-intercept. Line n is perpendicular to m and intersects line m at the first quadrant. The x-intercept of n must be

(A) Zero

(B) Greater than the x – intercept of m.

(C) Positive and less than the x – intercept of m.

(D) Negative and less than the x – intercept of m.

(E) Less than the x – intercept of m but it can be negative, zero or positive.

32. $(2\cos^2\theta - 5 + 2\sin^2\theta)^4$

(A) 0   (B) 1   (C) 16   (D) 81   (E) 256

33. Shown in figure 6 is a right prism whose bases are congruent isosceles trapezoids. Which of the given points is located in the plane determined by the vertices A, C and G?

(A) F

(B) B

(C) H

(D) D

(E) E

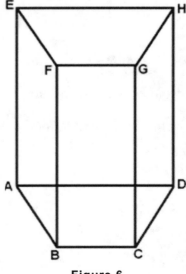

Figure 6

GO ON TO THE NEXT PAGE ▶▶▶

Model Test 4

34. Two identical squares are placed within a circle so that there is exactly one point common to both squares and each square intersects the circle at one and only one point so that when a point is selected on each square, the maximum distance between the points will be equal to the length of the diameter of the circle. What is the ratio of the area of the circle to the total area of the squares?

(A) 2   (B) 3   (C) π (D) 2π

(E) It cannot be determined from the information given.

35. In figure 7, triangles ABE and ADC are similar and z=18. What is the value of $\dfrac{x}{y}$?

(A) 1/3   (B) 3/4   (C) 5/12

(D) 4/3   (E) 12/5

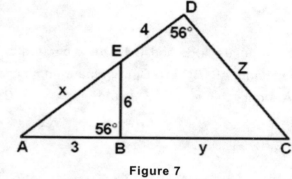

Figure 7

Figure not drawn to scale

36. What are the x-coordinates of the points of intersection of the line y = x and the circle having the radius of 5 and centered at (1,-2)?

(A) -2.7 and -3.7   (B) 2.7 and -3.7   (C) -2.7 and 3.7   (D) 2.7 and 3.7

(E) None of the above

37. At 08:00 AM, the amount of a radioactive element was 1,205 kg. If the amount decreases by 21% each hour, what will the amount of substance in kg's at 1:00PM?

(A) 0.49   (B) 370.79   (C) 469.35   (D) 2,583.02   (E) 3,125.46

GO ON TO THE NEXT PAGE ▶▶▶

38. In figure 8, if ΔABC is reflected across line l to get ΔA'B'C', what will be the sum of y – coordinates of vertices of ΔA'B'C'?

(A) -15　　　(B) 1　　　(C) 9　　　(D) 16

(E) The given information is not enough to solve this question.

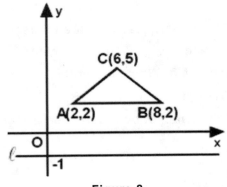

Figure 8

Figure not drawn to scale

39. If points A, B and C lie on a circle and if the center of the circle doesn't lie on either of the line segments AB, BC or AC, then the triangle can not be

(A) right　　　　(B) acute　　　　(C) obtuse　　　　(D) isosceles　　　　(E) equilateral

40. In figure 9, two identical cubes are placed side by side and each one has a side length of 2. What is the distance from vertex A to the midpoint C of BD?

(A) 6　　　(B) $\sqrt{21}$　　　(C) $2\sqrt{5}$　　　(D) $3\sqrt{2}$　　　(E) $\sqrt{7}$

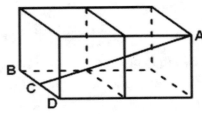

Figure 9

41. In figure 10, if $170° < a + c < 250°$, which of the following describes all possible values of b + d?

(A) $110° < b + d < 190°$　　　(D) $0° < b + d < 360°$

(B) $200° < b + d < 250°$　　　(E) $0° < b + d < 540°$

(C) $200° < b + d < 280°$

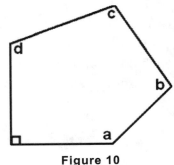

Figure 10

GO ON TO THE NEXT PAGE ▶▶▶

Model Test 4

42. If $f(x) = x^3 - 1$ and f^{-1} is the inverse of function f, what is $f^{-1}(3)$?

(A) $\dfrac{1}{26}$ (B) 1.33   (C) 1.59   (D) 26   (E) 28

43. $f(x) = x^3 + x + 1 \Rightarrow (f \circ f^{-1})(2) = ?$

(A) 11   (B) $\dfrac{1}{2}$ (C) $\dfrac{1}{11}$ (D) 2   (E) It cannot be determined because f(x) is not invertible.

44. For the cube given in figure 11, E is the midpoint of the edge KL. If each side of the cube is 4 inches, what is the perimeter of the triangle DEC?

(A) 10   (B) 11   (C) 12   (D) 16   (E) 18

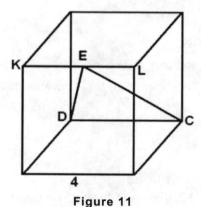

Figure 11

45. The n - th term in the sequence of numbers $\{15, 19, 23, 27, 31, ...\}$ is

(A) 4n+15   (B) 4n+11   (C) n+14   (D) 4n   (E) n+4

46. Two positive integers a, and b satisfy the relation a △ b, if and only if a = 2b + 1. If a, b, and c satisfy the relations a △ b and b △ c, what is the value of a in terms of c?

(A) 2c + 1   (B) 2c + 3   (C) 4c + 1   (D) 4c + 3

(E) It cannot be determined from the information given.

GO ON TO THE NEXT PAGE ▶▶▶

Model Test 4

47. If the area bounded by the x–axis and the upper half of the circle with equation $(x-4)^2+(y+1)^2=9$ is A, what is the area bounded by the x–axis and the upper half of the circle with equation $(x+4)^2+(y+1)^2=9$?

(A) A/2   (B) A   (C) 2A   (D) A^2 (E) 4A

48. Figure 12 shows two points A and B on a cylinder with radius of length 2.5 inches and altitude of length 12 inches. If points A and B are joined by a curved path of minimum length as indicated in the figure, what is the length in feet of this path?

(A) 1.19   (B) 1.20   (C) 13.4   (D) 14.3

(E) It cannot be determined from the information given.

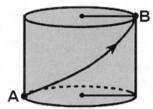

Figure 12

49. $196 = A + B\cdot3 + C\cdot9 + D\cdot27 + E\cdot81$

If each of the numbers A, B, C, D, and E is a member of the set {0, 1, 2} then what is the remainder when the five digit number ED,CBA is divided by 4?

(A) 0   (B) 1   (C) 2   (D) 3   (E) 4

50. The rectangular box given in figure 13 has a base with sides of length 2x and 3x and a diagonal of length 7x. What is the height of the prism?

(A) 3x   (B) 4x   (C) 5x   (D) 6x

(E) It cannot be obtained from the given information.

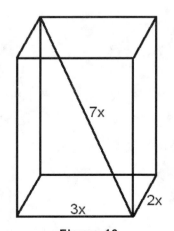

Figure 13

Figure not drawn to scale

S T O P

END OF TEST

(Answers on page 201 – Solutions on page 211)

Model Test 5

Test Duration: 60 Minutes

Directions: For each of the following problems, decide which is the **best** of the choices given. If the exact numerical value is not one of the choices, select the choice that best approximates this value. Then fill in the corresponding oval on the answer sheet.

Notes:

- A calculator will be necessary for answering some (but not all) of the questions in this test. For each question you will have to decide whether or not you should use a calculator. The calculator you use must be at least a scientific calculator; programmable calculators and calculators that can display graphs are permitted.

- The only angle measure used on this test is degree measure. Make sure your calculator is in the degree mode.

- Figures that accompany problems in this test are intended to provide information useful in solving the problems. They are drawn as accurately as possible **except** when it is stated in a specific problem that its figure is not drawn to scale.

- All figures lie in a plane unless otherwise indicated.

- Unless otherwise specified, the domain of any function f is assumed to be the set of all real numbers **x** for which **f(x)** is a real number.

Reference Information: The following information is for your reference in answering some of the questions in this test.

- Volume of a right circular cone with radius **r** and height **h**: $V = \frac{1}{3}\pi r^2 h$

- Lateral area of a right circular cone with circumference of the base **c** and slant height **l**: $S = \frac{1}{2}cl$

- Volume of a sphere with radius **r**: $V = \frac{4}{3}\pi r^3$

- Surface area of sphere with radius **r**: $S = 4\pi r^2$

- Volume of a pyramid with base area **B** and height **h**: $V = \frac{1}{3}Bh$

1. $\sqrt{14^2 - 13^2} = 3^n$, n=

(A) 0.5   (B) 1   (C) 1.5   (D) 2   (E) 2.5

2. $\sqrt{\dfrac{\left(\frac{1}{5}\right)^{-1} : \left(\frac{1}{5}\right)^2}{5^{-3}}} = ?$

(A) 1/125   (B) 1/25   (C) 1   (D) 25   (E) 125

GO ON TO THE NEXT PAGE ▶▶▶

Model Test 5

3. For a set of 11 consecutive positive integers, which of the following must be correct?

    I. Median is a positive integer.

    II. Median equals mean.

    III. There is no mode in this set.

(A) I only   (B) II only   (C) III only   (D) I and II only   (E) I, II, and III.

4. When $64.8 \cdot 10^{-6}$ is written in the decimal form, what is the number of zeros between the decimal point and the first nonzero digit counted from left?

(A) 8   (B) 6   (C) 5   (D) 4   (E) 3

5. If $-1 < a < b < 0$ and $x = a \cdot b$ then which of the following graphs can most likely demonstrate the position of x on the number line?

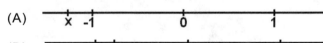

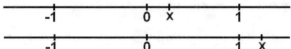

6. Which of the following is not correct for the function defined by $f(x) = x^4$?

(A) $f(a^n) = (f(a))^n$ (B) $f(-a) = f(a)$ (C) $f\left(\dfrac{a}{b}\right) = \dfrac{f(a)}{f(b)}$ (D) $f\left(\dfrac{1}{b}\right) = \dfrac{1}{f(b)}$ (E) $f(a + b) = f(a) + f(b)$

7. Which of the following is not in the solution set of $5x - 2y < 21$?

(A) (3, 4)   (B) (6, 6)   (C) (-3, 1)   (D) (4, -2)   (E) (2, 3)

GO ON TO THE NEXT PAGE ▶▶▶

Model Test 5

8. A certain brand of wedding ring is made in such a way that 20% of its price is because of the gold and the rest is because of the silver used in producing the ring. The price of gold increases by 12% every year, however the price of silver stays the same. To the nearest dollar, what will be the price of the ring 3 years later if it costs $100 at present?

(A) $100   (B) $108   (C) $132   (D) $140

(E) None of the above

9. If $f(A) = \dfrac{(A^7)^2}{A^8}$ then $f(3.3) = ?$

(A) 391.3   (B) 391.4   (C) 1291.4   (D) 1291.5   (E) 46411.5

10. If b and $-4a$ are consecutive integers, which of the following is always odd?

(A) a + b   (B) a - b   (C) a + 2b   (D) a − 2b   (E) 2a + b

11. In the formula $E = mc^2$, the effect of doubling the value of c is to

(A) Multiply the value of E by 4   (B) Double the value of E

(C) Multiply the value of E by 0.5   (D) Multiply the value of E by 0.25

(E) None of the above

12. If in figure 1, the tangent of angle ACB is 2 then what is the length of BC?

(A) 3   (B) 6   (C) 9   (D) 12

(E) None of the above

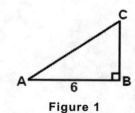

Figure 1

Figure not drawn to scale

GO ON TO THE NEXT PAGE ▶▶▶

13. The solution set of the equation given by $\sqrt{x-\frac{2}{3}}+\sqrt{x-\frac{3}{4}}=\sqrt{x-\frac{3}{2}}+\sqrt{x-\frac{4}{3}}$ consists of

(A) no elements　　　(B) 1 element　　　(C) 2 elements　　　(D) 3 elements　　　(E) 4 elements

14. Three less than twice x cubed equals one more than half of y squared. Which of the following correctly represents the relation between x and y?

(A) $2x^3 - 3 = 1 + \dfrac{y^2}{2}$　　　(B) $2x^3 - 3 = 1 + 2y^2$　　　(C) $3 - 2x^3 = 1 + \dfrac{y^2}{2}$

(D) $2x^3 + 3 = 1 + \dfrac{y^2}{2}$　　　(E) $2x^3 + 3 = 1 + 2y^2$

15. Which of the following describes the interior region of the circle with center (4,3) and radius 2?

(A) $(x+4)^2 + (y+3)^2 \le 4$　　　(B) $(x-4)^2 + (y-3)^2 \le 4$　　　(C) $(x-4)^2 + (y-3)^2 < 4$

(D) $(x+4)^2 + (y+3)^2 < 4$　　　(E) $(x-4)^2 + (y-3)^2 > 4$

16. It is given in figure 2 that OABC is a square, In order to calculate the area of the shaded region which of the following must be known?

(A) Coordinates of B and G　　　(B) Coordinates of F and G

(C) Coordinates of C and F　　　(D) Coordinates of B and E

(E) Coordinates of D and G

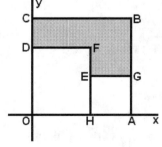

Figure 2

17. If $2 \cdot \sin^2 x - 3 \cdot \sin x = 1 - 2 \cdot \cos^2 x$ then sinx = ?

(A) 1　　　(B) $-\dfrac{1}{3}$　　　(C) $\dfrac{2}{3}$　　　(D) $\dfrac{1}{3}$　　　(E) $\dfrac{1}{9}$

GO ON TO THE NEXT PAGE ▶▶▶

Model Test 5

18. A unitary number is one such that when its digits are added, when the digits of the resulting number are added, and so on, the final result is 1. For example, 8767 is a unitary number as 8+7+6+7=28; 2+8=10; and 1+0=1. If k is a unitary number which of the following is the next greater unitary number?

(A) k − 9   (B) k + 3   (C) k + 6   (D) k + 9   (E) 9k

19. If the length of the longest chord of that can be drawn within a circle C_1 equals the radius of another circle C_2, what is the ratio of the area of C_1 to that of C_2?

(A) 4   (B) 2   (C) 1   (D) 1 / 2   (E) 1 / 4

20. What is the domain of the relation given in figure 3?

(A) {-2, 1, 2, 3}

(B) -6 < x < 5

(C) -6 ≤ x ≤ 5

(D) -6 ≤ x < -3 and -2 ≤ x < 5

(E) -6 ≤ x < -3 or -2 ≤ x < 5

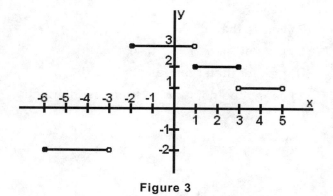

Figure 3

21. If $2 < x^2 < 14$ then x cannot be

(A) -π   (B) $\sqrt{7}$ (C) $\sqrt[3]{15}$ (D) $-\sqrt{5}-\sqrt{3}$ (E) $\sqrt{12}-1$

22. It is given that two complementary angles have a ratio of 5:13. What is the measure of the greater angle?

(A) 5°   (B) 25°   (C) 35°   (D) 55°   (E) 65°

GO ON TO THE NEXT PAGE ▶▶▶

Model Test 5

23. In figure 4, circles P and Q are tangent to the line m at points A and B respectively. If the radii of the circles are 2 and 4 inches respectively, and the centers of the circles are 10 inches apart, what is the length of the segment AB in inches?

(A) 5   (B) 6   (C) 7   (D) 8

(E) None of the above

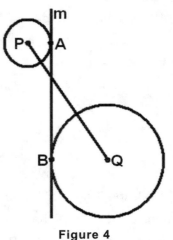

Figure 4

Figure not drawn to scale

24. In figure 5, a circle with radius r is given. If the parallel chords AB and CD are equal in length, then, the length of segment AC

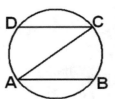

(A) equals 2r.

(B) is less than r.

(C) greater than 2r.

(D) greater than r but less than 2r.

(E) cannot be determined from the information given.

Figure 5

25. The volume V of a circular cone of radius r and altitude h is given by the following formula:

$V = \dfrac{1}{3}\pi r^2 h$; which of the following statements is true?

(A) If h is doubled, V is quadrupled.

(B) If r is tripled, V is multiplied by 9.

(C) If both r and h are doubled, V is doubled.

(D) If r is doubled, and h is halved, V does not change.

(E) If r is halved, and h is doubled, V does not change.

GO ON TO THE NEXT PAGE ▶ ▶ ▶

Model Test 5

26. The parametric equations given by $x = -t^2$ and $y = 2t^2 + 1$ represent

(A) a parabola

(B) portion of a parabola

(C) portion of a line

(D) a line

(E) portion of a hyperbola

27. In figure 6, for all triangles ABC

(A) $m \angle \theta < m \angle \beta$ (B) $m \angle \theta < m \angle B$ (C) $m \angle \theta > m \angle \beta$

(D) $m \angle C < m \angle B$ (E) $m \angle \theta < m \angle \alpha$

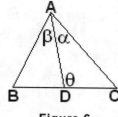

Figure 6

28. In figure 7, which pair of points can be joined to give a line with negative slope?

(A) A and E   (B) A and D   (C) B and D   (D) B and C   (E) C and D

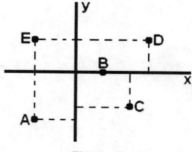

Figure 7

29. How many integers are there in the solution set of $|3x - 6| + 2 < 0$?

(A) None   (B) 3   (C) 4   (D) 5   (E) More than 5

30. For every real value of x, the equation given by $4x^2 - 6x + 2c = (2x - 4)(ax - b)$ is satisfied. What is the value of $a - b - c$?

(A) 3   (B) 4   (C) 5   (D) 6   (E) 7

GO ON TO THE NEXT PAGE ▶▶▶

Model Test 5

31. How many digits is the number 55^{22}?

(A) 22   (B) 23   (C) 37   (D) 38   (E) 39

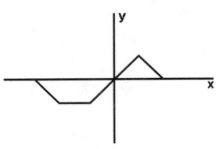

32. Graph of the relation given above is reflected across the origin. What will be the resulting graph?

(A)   (B)   (C)

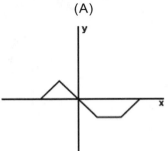
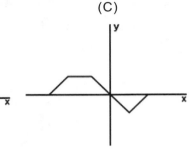

(D)   (E)

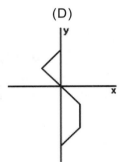
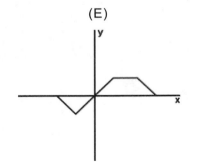

33. In *RUSH* Academy of 2,200 students, 800 are studying Spanish, 700 are studying French, and 300 are studying German. What is the greatest possible number of these students that might NOT be studying any of these languages if at least one of the students studies more than one language?

(A) 400   (B) 401   (C) 1400   (D) 1500   (E) 1900

34. If the lines $x - 2y = 10$ and $ax + y = 5$ are perpendicular to each other then $a =$

(A) -2   (B) -1   (C) 0   (D) 1   (E) 2

GO ON TO THE NEXT PAGE ▶▶▶

Model Test 5

	Chemistry	Physics	Biology	Total
French				21
Spanish	4		12	
Total	11	8		40

35. Each student in a class of 40, study one foreign language and one branch of science and their choices are shown in the partially completed table above. What is the probability that a randomly chosen student studies physics given that the student studies French?

(A) 5/21   (B) 5/8   (C) 5/4   (D) 24/40

(E) None of the above

36. Based on the data given in figure 8 what is the perimeter of the shaded rectangle?

 (A) 22   (B) 25   (C) 28   (D) 34   (E) 50

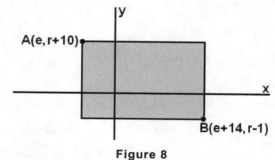

Figure 8

Figure not drawn to scale

37. Sequential arrangements of congruent isosceles right triangles with legs of unit length are formed according to a pattern given in figure 9. Each arrangement after the first is formed by adding another layer of triangles below the previous arrangement as shown. If this pattern continues, what will be the number of unit line segments used in the 4th arrangement?

(A) 10   (B) 16   (C) 20   (D) 24   (E) 30

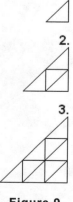

Figure 9

38. If $x = 2$, $y = -1$ and $z = 1$, the expression given by $\dfrac{a^2 - xbc - xac + yb^2}{a + b \cdot z}$ simplifies to

GO ON TO THE NEXT PAGE ▶▶▶

Model Test 5

(A) a − b − c   (B) a + b + c   (C) a + b + 2c   (D) a − b − 2c   (E) a − b + 2c

39. In figure 10, l_1 // l_2. The bisectors of angles AKL and BML (not shown) intersect at how many degrees?

(A) 10°   (B) 15°   (C) 20°   (D) 30°   (E) 40°

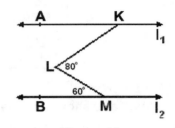

Figure 10

Figure not drawn to scale

40. A bacteria population that has 128 inhabitants initially doubles every hour. How many inhabitants will there be after 12 hours?

(A) 2^{12} (B) 2^{15} (C) 2^{18} (D) 2^{19} (E) 2^{20}

41. If $a^4 < b^4$, which of the following must be true?

     I. a < b   II. $a^2 < b^2$ III. $|a| < |b|$

(A) I only   (B) II only   (C) III only   (D) II and III only   (E) I, II, and III

42. How many of the functions given have a positive y intercept?

(A) 1   (B) 2   (C) 3   (D) 4   (E) 5

$y_1 = (3-x)(x-5)$

$y_2 = -x \cdot (4+x)$

$y_3 = (2-x) \cdot (4-2x)$

$y_4 = \dfrac{1-x}{x+2}$

$y_5 = 3^{2-x}$

43. Circle C is given by the equation $(x + 3)^2 + (y − 4)^2 = 25$. If (1, 7) is one endpoint of a diameter of this circle, then the other endpoint is

(A) (-7, 1)   (B) (7, -1)   (C) (5, 10)   (D) (6, 12)   (E) (- 4, 2)

GO ON TO THE NEXT PAGE ▶▶▶

Model Test 5

44. In **RUSH** Publications there are five test controllers Alihan, Berk, Cemal, Deniz, and Enis whose working rates can be calculated using the information given by the graph in figure 11. Which of the following is a correct statement if the lettered points represent the initials of the controllers?

(A) Berk is the slowest of them all.

(B) Deniz is the fastest of them all.

(C) Deniz and Enis work at the same rate.

(D) Alihan and Berk work at the same rate.

(E) Cemal and Enis work at the same rate.

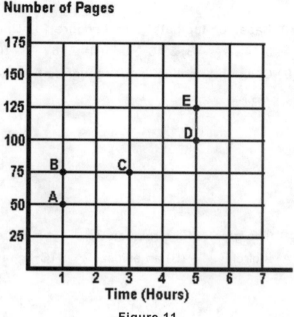

Number of Pages

Time (Hours)

Figure 11

45. If $\sin\theta = 0.38$, then $\sin(90° + \theta) = ?$

(A) -0.92   (B) -0.38   (C) 0.38   (D) 0.92

(E) None of the above

46. What is the approximate radius of the circle that passes through the vertices of the isosceles triangle given in figure 12?

(A) 5   (B) 6   (C) 7   (D) 8   (E) 9

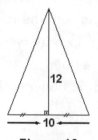

Figure 12

Figure not drawn to scale

47. Which of the following equals the reciprocal of the fraction given by $\dfrac{a^{-1}+b^{-1}}{a^{-1}-b^{-1}}$?

(A) $\dfrac{a+b}{a}$ (B) $\dfrac{a+b}{b}$ (C) $\dfrac{a+b}{a-b}$ (D) $\dfrac{b+a}{b-a}$ (E) $\dfrac{b-a}{b+a}$

GO ON TO THE NEXT PAGE ▶▶▶

48. Based on the data given in figure 13, what is the length of segment AB?

(A) 3.14   (B) 3.46   (C) 5.66   (D) 6.28

(E) Given information is not enough to find the length of AB.

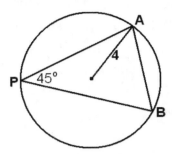

Figure 13

Figure not drawn to scale

49. The graph of the relation $9x^2 + 4y^2 = 36$ represents

(A) a circle   (B) an ellipse   (C) a parabola   (D) a hyperbola   (E) a straight line

50. The circle given in figure 14.A is segmented into 10 non overlapping sectors colored in either black or white and each sector is partitioned into 5 disjoint sections making up a total of 50 non overlapping regions. The regions in figure 14.A will be re-colored so that in each sector exactly one region will be gray and all other regions will be white with the restriction that no two regions colored in gray will have a common boundary. Thus, a total of 10 gray and 40 white regions

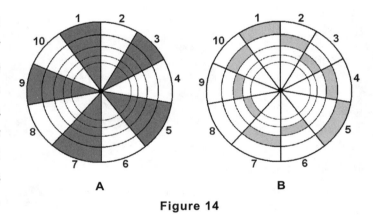

Figure 14

should be obtained and one example is shown in figure 14.B. In how many different ways can this be done?

(A) 5^{10} (B) $5 \cdot 4^9$ (C) $15 \cdot 4^8$ (D) P(50, 5)   (E) C(50, 5)

S T O P

END OF TEST

(Answers on page 201 – Solutions on page 213)

Model Test 6

Test Duration: 60 Minutes

Directions: For each of the following problems, decide which is the **best** of the choices given. If the exact numerical value is not one of the choices, select the choice that best approximates this value. Then fill in the corresponding oval on the answer sheet.

Notes:

- A calculator will be necessary for answering some (but not all) of the questions in this test. For each question you will have to decide whether or not you should use a calculator. The calculator you use must be at least a scientific calculator; programmable calculators and calculators that can display graphs are permitted.

- The only angle measure used on this test is degree measure. Make sure your calculator is in the degree mode.

- Figures that accompany problems in this test are intended to provide information useful in solving the problems. They are drawn as accurately as possible **except** when it is stated in a specific problem that its figure is not drawn to scale.

- All figures lie in a plane unless otherwise indicated.

- Unless otherwise specified, the domain of any function f is assumed to be the set of all real numbers **x** for which **f(x)** is a real number.

Reference Information: The following information is for your reference in answering some of the questions in this test.

- Volume of a right circular cone with radius **r** and height **h**: $V = \frac{1}{3}\pi r^2 h$

- Lateral area of a right circular cone with circumference of the base **c** and slant height **l**: $S = \frac{1}{2}cl$

- Volume of a sphere with radius **r**: $V = \frac{4}{3}\pi r^3$

- Surface area of sphere with radius **r**: $S = 4\pi r^2$

- Volume of a pyramid with base area **B** and height **h**: $V = \frac{1}{3}Bh$

1. Which of the following is an irrational number?

(A) $\dfrac{2\sqrt{3}}{\sqrt{12}}$ (B) $\sqrt{\dfrac{1}{2}} \cdot \sqrt{18}$ (C) $\dfrac{3\sqrt{10}}{2\sqrt{2}}$ (D) $\sqrt[3]{-8}$ (E) $\sqrt{3}(3\sqrt{3} - \sqrt{27})$

2. $2^{x-1} = A$ and $A = \dfrac{1}{4}$; x=?

(A) -1   (B) -2   (C) 0   (D) 1   (E) 2

GO ON TO THE NEXT PAGE ▶▶▶

Model Test 6

3. The midpoint of the line segment joining the points (-3, 4) and (3,-4) is

(A) (-4,-4) (B) (-4,3) (C) (4,3) (D) (0,0) (E) (3,3)

4. If $f(x) = x^3 - 1$ and $g(x) = (x - 1)^3$ then $f(g(2)) - g(f(2))$?

(A) 0 (B) 7 (C) 276 (D) 512 (E) -216

5. What is the multiplicative inverse of $-1 + \sqrt{3}$?

(A) $-1 - \sqrt{3}$ (B) $1 - \sqrt{3}$ (C) $1 + \sqrt{3}$ (D) $\sqrt{3}$ (E) $\dfrac{1+\sqrt{3}}{2}$

6. In a group of 26 students, 18 play football, 12 play basketball and 4 play neither game. How many students play both of the games?

(A) no one (B) 4 (C) 6 (D) 8 (E) 12

7. If x, y, and z are nonzero real numbers and if $\dfrac{x + y^3}{z^2} = zy^3$ then x =

(A) z^3 (B) $2y^3$ (C) $z^3 + 2y^3$ (D) $y^3(z^3 - 1)$ (E) $y^3(z^3 + 1)$

8. How many distinct even numbers are there between 250 and 400, inclusive?

(A) 73 (B) 74 (C) 75 (D) 76 (E) 77

GO ON TO THE NEXT PAGE ▶▶▶

Model Test 6

9. $x - y = 2$; $z = 3$; $4y - 4x + 2z = ?$

(A) -2　　　　　　(B) -8　　　　　　(C) 2　　　　　　(D) 6　　　　　　(E) 14

10. If $x^* = x^5 - 1$, then the value of $(-1)^* - 1^* =$

(A) -8　　　　　　(B) -4　　　　　　(C) -2　　　　　　(D) 0　　　　　　(E) 2

11. The equation given by $(x^2 + y^2 - 1)(\sqrt{x^2 + y^2} - 2) = 0$ represents

(A) the empty set.

(B) two externally tangent circles.

(C) two concentric circles with distinct radii.

(D) two circles of different size intersecting at two points.

(E) two circles of the same size intersecting at two points.

12. The total cost, in dollars, of a telephone call that is m minutes in length from City R to City M is given by the function $f(m) = 2.34 (0.65 \lceil m \rceil + 1)$, where m>0 and $\lceil m \rceil$ is the least integer greater than or equal to m. What is the total cost of a 5.5 minute telephone call from City R to City M?

(A) 9 dollars and 94 cents　　　(B) 9 dollars and 95 cents　　　(C) 10 dollars and 71 cents

(D) 11 dollars and 46 cents　　　(E) 11 dollars and 47 cents

13. If tanA=0.9511 and sinA=0.3090, what is the value of cosA?

(A) 0.325　　　　(B) 0.294　　　　(D) 3.078　　　　(E) 3.078

(E) It cannot be determined from the information given.

GO ON TO THE NEXT PAGE ▶▶▶

Model Test 6

14. What is the approximate slope of the line $\sqrt{5}x + 5y - 2\sqrt{5} = 0$?

(A) -2.45   (B) -2.24   (C) -1.24   (D) -0.44   (E) -0.45

15. If a and b are nonzero real numbers and $(4.12)^a = (8.26)^b$, what is the value of $\left(\dfrac{a}{b}\right)^2$?

(A) 0.50   (B) 1.49   (C) 2.20   (D) 2.22   (E) 4.40

16. For a set of 12 consecutive positive even integers, which of the following must be correct?

    I. Median is a positive odd integer.

    II. Median equals mean.

    III. There is no mode in this set.

(A) I only   (B) II only   (C) III only   (D) I and II only   (E) I, II, and III.

17. If p and q are distinct prime numbers greater than 2, which of the following can be prime?

(A) p^4 (B) pq   (C) 5q   (D) p + q   (E) 2p – q

18. Let ABC be a triangle with sides of integral length in the ratio 3:4:5. Which of the following can not be the area of $\triangle$ ABC?

(A) 6   (B) 24   (C) 40   (D) 54   (E) 96

GO ON TO THE NEXT PAGE ▶▶▶

Model Test 6

19. In a classroom there are 10 boys including Tibet and Emre and 8 girls including Sevda. Tibet will give a party and he wants to invite 5 girls and 3 boys to the party. If he plans to invite Sevda but not Emre, in how many different ways can he make the selection?

(A) $\binom{10}{5} \cdot \binom{8}{3}$ (B) $\binom{8}{3} \cdot \binom{7}{4}$ (C) $\binom{7}{3} \cdot \binom{8}{4}$ (D) $\binom{9}{3} \cdot \binom{7}{4}$ (E) $\binom{9}{4} \cdot \binom{7}{3}$

20. If a proposition is given by $p \Rightarrow q$ then $q' \Rightarrow p'$ is

(A) the inverse of this proposition   (B) the negation of this proposition

(C) the converse of this proposition   (D) the contrapositive of this proposition

(E) None of the above.

21. In figure 1, AB // CD. What is the value of sin(x)?

(A) $\cos \dfrac{\theta}{2}$ (B) $-\cos \dfrac{\theta}{2}$ (C) $\sin \dfrac{\theta}{2}$ (D) $-\sin \dfrac{\theta}{2}$

(E) None of the above.

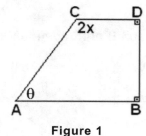

Figure 1

22. If $f(x) = x^2 - 12$ then $f^{-1}(x) = ?$

(A) $x^2 - 12$ (B) $\sqrt{x+12}$ (C) $-x^2 + 12$ (D) $\pm\sqrt{x+12}$ (E) $\dfrac{1}{x^2 - 12}$

GO ON TO THE NEXT PAGE ▶▶▶

Model Test 6

23. A right circular cylinder is given such that the base radius is 6 and the height is 9. If A is a point on the upper base and B is a point on the lower base of the cylinder, what is the difference between the max and the min values of AB?

(A) 6   (B) 9   (C) 12   (D) 15

(E) None of the above.

24. If f(g(x))= 3x-1 and g(x)= $\frac{x-1}{2}$, which of the following is f(x)?

(A) $\frac{3x+1}{2}$ (B) $\frac{3x-1}{2}$ (C) 6x   (D) 2(3x+1)   (E) 2(x+3)

25. The probability of event A occurring is $\frac{1}{3}$ and the probability of event B occurring is $\frac{1}{4}$. Events A and B are dependent on each other and the probability of events A and B occurring simultaneously is $\frac{1}{8}$. What is the probability of event A or B occurring?

(A) 0   (B) $\frac{11}{24}$ (C) $\frac{13}{24}$ (D) $\frac{15}{24}$ (E) 1

26. Under what conditions would ac = bc be true?

(A) a = b   (B) c = 0   (C) a = b and c = 0   (D) a = b or c = 0

(E) None of the above.

27. If the 5th term of an arithmetic sequence is 12 and the 50th term of the sequence is 102, what is the first term of the sequence?

(A) 1   (B) 2   (C) 3   (D) 4   (E) 5

GO ON TO THE NEXT PAGE ▶ ▶ ▶

Model Test 6

28. Given that $f(x) = ax^2 + bx + c$. If $f(x)$ and $|f(x)|$ intersect at no points then which of the following statements must be correct?

(A) $\dfrac{c}{a} < 0$ and $\dfrac{b}{a} > 0$

(B) $b^2 - 4ac > 0$ and $a < 0$

(C) $b^2 - 4ac < 0$ and $a < 0$

(D) $\dfrac{b}{2a} > 0$ and $a < 0$

(E) $b^2 - 4ac < 0$ and $a > 0$

29. If the lines given by 3x + 2y = 12 and 6x − Ay = 11 are parallel then A =

(A) 4   (B) 3   (C) 2   (D) − 2   (E) − 4

30. If x, y and z are the angles of a triangle and one third of the sum of the measures of x and y is 35° then z measures how many degrees?

(A) 70°   (B) 75°   (C) 105°   (D) 110°   (E) 145°

31. If $4x + 1 > 3^x$ then x can be

(A) -2   (B) -1   (C) 0   (D) 1   (E) 2

32. If x is a real number for which |x − 1| < 4 then which of the following cannot be concluded?

(A) -7 < x < 7   (B) |x| < 7   (C) |x| < 5   (D) -3 < x < 5   (E) |x| < 2

GO ON TO THE NEXT PAGE ▶▶▶

Model Test 6

33. What is the sum of all interior and exterior angles of a regular heptagon (7 sides)?

(A) $5 \cdot 180°$ (B) $6 \cdot 180°$ (C) $7 \cdot 180°$ (D) $8 \cdot 180°$ (E) $9 \cdot 180°$

34. Graph of $f(x)$ is given in the figure 2. What is the graph of $f(x-1)+1$?

(A)

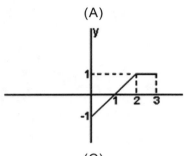

(B)

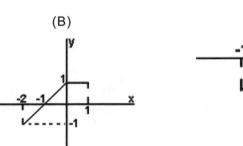

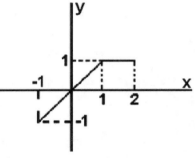

Figure 2

(C)

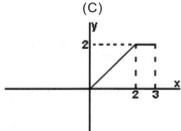

(D)

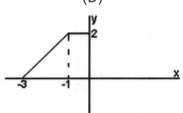

(E)

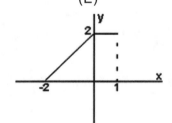

35. If $\log_2 x^3 = 15$ then $x = ?$

(A) 3 (B) 5 (C) 25 (D) 32

(E) None of the above.

36. A parabola given by $f(x) = ax^2 + bx + c$ passes through points (1, 4), (0, 3), and (-1, 6). What is the equation of this parabola?

(A) $y = -2x^2 - x + 3$ (B) $y = 2x^2 + x - 3$ (C) $y = x^2 - 2x + 3$

(D) $y = 2x^2 - 3x + 1$ (E) $y = 2x^2 - x + 3$

GO ON TO THE NEXT PAGE ▶▶▶

Model Test 6

37. A motorboat having a maximum speed of 20 mph in still water; is in a 4 mph stream. If it travels 48 mi downstream and the same distance back at up steam, then what is its average rate for the round trip in mph?

(A) 9.6   (B) 16.0   (C) 19.2   (D) 20.0   (E) 24.0

38. Graph of f(x) is given in figure 3. What will be the graph of -f(-x)?

(A)
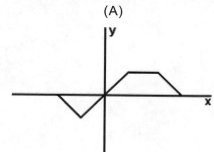

(B)
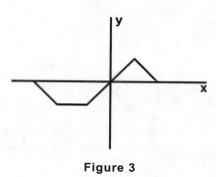

Figure 3

(C)
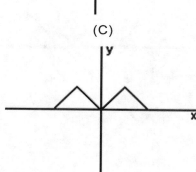

(D)
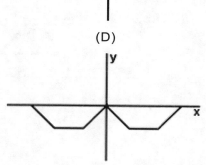

(E)
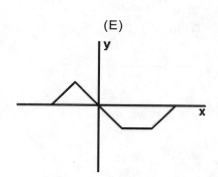

39. A rectangular garden with dimensions a and b feet is bordered by a walk by extending each edge of the garden by w feet as shown in figure 4. What is the area of the walk given by the shaded region?

(A) (a + w) (b + w)

(B) (a + w) (b + w) − ab

(C) (a + 2w) (b + 2w)

(D) (a + 2w) (b + 2w) − ab

(E) ab − (a − 2w) (b − 2w)

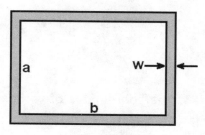

Figure 4

GO ON TO THE NEXT PAGE ▶▶▶

Model Test 6

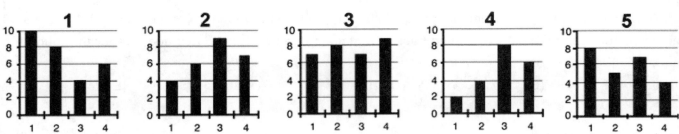

40. Five different groups of data are represented by the graphs given above such that in each group four different data are presented. For example in data group 1 there are 4 numbers: 10, 8, 4 and 6. For which group is the mean greater than the median?

(A) 1   (B) 2   (C) 3   (D) 4   (E) 5

41. What is the domain of the function given in figure 5?

(A) $(a, e) - \{g\}$ (B) (a, e) (C) $[c, e)$

(D) $[d, f) - \{h\}$ (E) $[d, f)$

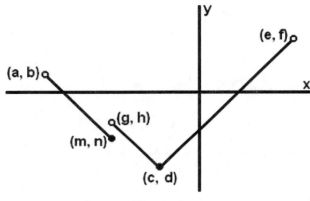

Figure 5

42. If $(a + 2b)^2 \geq (a - 2b)^2 + 18$ where a and b are positive real numbers and a cannot be greater than b then what is the least possible value of b?

(A) 1.3   (B) 1.5   (C) 1.8   (D) 2.1   (E) It cannot be determined from the information given.

43. The coordinates of vertices A and B of an isosceles triangle ABC are (3,5) and (3,17) respectively. If AC = BC, then which of the following cannot be determined?

(A) Length of the base   (B) Length of the altitude to the base

(C) Midpoint of the base   (D) Equation of the median to the base

(E) Equation of the base

GO ON TO THE NEXT PAGE ▶ ▶ ▶

For questions 44 – 45 please refer to the data based on the employment in a certain county given in the following graph:

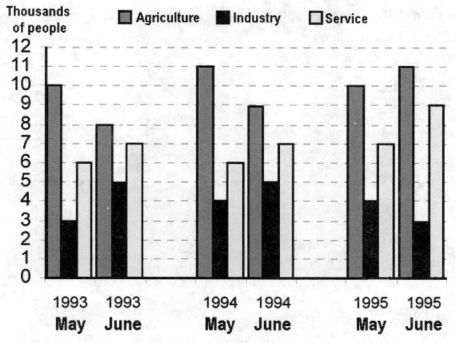

44. In May 1994, how many more people were employed in agriculture than in industry?

(A) 3000   (B) 4000   (C) 6000   (D) 7000   (E) 8000

45. What was the percent increase in the average number of people employed in the service sector in the May – June period from 1993 to 1995?

(A) 14%   (B) 17%   (C) 18%   (D) 19%   (E) 23%

46. What is the diameter in inches of the circle that is inscribed in a triangle with sides of length 3, 4 and 5 all in feet?

(A) 24   (B) 12   (C) 2   (D) 1

(E) None of the above.

GO ON TO THE NEXT PAGE ▶ ▶ ▶

Model Test 6

47. If AB || DE in figure 6 then which of the following is a correct statement?

    I. Triangles ABC and EDC are congruent.

    II. Triangles ABC and EDC are similar.

    III. Triangles ABC and EDC are isosceles.

(A) I only   (B) I and III only

(C) II only   (D) I, II and III.

(E) III only

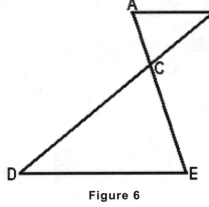

Figure 6
Figure not drawn to scale.

48. For the right triangle given in figure 7, $\sin A + \tan A = \dfrac{32}{15}$. What is the length of AB?

 (A) 2.5   (B) 3.0   (C) 3.5   (D) 5.0

(E) None of the above.

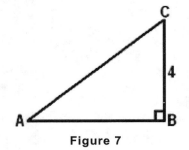

Figure 7

49. For a function given by $f(x) = x - \dfrac{1}{x}$ which of the following are equal in value?

I. $-f(x)$ II. $f(-x)$ III. $f\left(\dfrac{-1}{x}\right)$ IV. $f\left(\dfrac{1}{x}\right)$

(A) I and II only   (B) I and IV only   (C) I, II and III only

(D) I, II and IV only   (E) all are equal in value

50. If the polynomial $P(x) = x^3 - 2x^2 + 3px - 14$ is divisible by $(x+2)$ then $p = ?$

(A) -16   (B) -8   (C) -5   (D) 5   (E) 8

S T O P

END OF TEST

(Answers on page 201 – Solutions on page 215)

Model Test 7

Test Duration: 60 Minutes

Directions: For each of the following problems, decide which is the **best** of the choices given. If the exact numerical value is not one of the choices, select the choice that best approximates this value. Then fill in the corresponding oval on the answer sheet.

Notes:

- A calculator will be necessary for answering some (but not all) of the questions in this test. For each question you will have to decide whether or not you should use a calculator. The calculator you use must be at least a scientific calculator; programmable calculators and calculators that can display graphs are permitted.

- The only angle measure used on this test is degree measure. Make sure your calculator is in the degree mode.

- figures that accompany problems in this test are intended to provide information useful in solving the problems. They are drawn as accurately as possible **except** when it is stated in a specific problem that its figure is not drawn to scale.

- All figures lie in a plane unless otherwise indicated.

- Unless otherwise specified, the domain of any function f is assumed to be the set of all real numbers **x** for which **f(x)** is a real number.

Reference Information: The following information is for your reference in answering some of the questions in this test.

- Volume of a right circular cone with radius **r** and height **h**: $V = \frac{1}{3}\pi r^2 h$

- Lateral area of a right circular cone with circumference of the base **c** and slant height **l**: $S = \frac{1}{2}cl$

- Volume of a sphere with radius **r**: $V = \frac{4}{3}\pi r^3$

- Surface area of sphere with radius **r**: $S = 4\pi r^2$

- Volume of a pyramid with base area **B** and height **h**: $V = \frac{1}{3}Bh$

1. If $\frac{1}{t+3} = 6$, then $\frac{1}{t-3} =$

(A) -5/6　　　　(B) -6/5　　　　(C) -6/35　　　　(D) -35/6　　　　(E) 0

2. A club of 120 members must vote to elect one of the two candidates. If 80% of the decided members support Ferhat, and 18 members support Serhat, then how many club members are undecided?

(A) 24　　　　(B) 30　　　　(C) 40　　　　(D) 80　　　　(E) 96

GO ON TO THE NEXT PAGE ▶▶▶

Model Test 7

3. Sinem traveled to Bodrum. She traveled $\frac{1}{7}$ of the way with a bus, then walked for 6 miles, and then took a train for $\frac{5}{7}$ of the way, arriving Bodrum. How many miles did she travel altogether?

(A) 12   (B) 18   (C) 15   (D) 30   (E) 42

4. The reciprocal of $5 - 2\sqrt{3}$ is equal to

(A) $\dfrac{1}{-5 + 2\sqrt{3}}$ (B) $\dfrac{1}{5 + 2\sqrt{3}}$ (C) $\sqrt{13}$ (D) $5 + 2\sqrt{3}$ (E) $\dfrac{5 + 2\sqrt{3}}{13}$

5. The product of five integers is a negative odd number. Which of the following can be true of the five numbers?

(A) All five are positive and odd.

(B) All five are negative and even.

(C) Two are positive odd and three are negative odd.

(D) Two are positive even and three are negative odd.

(E) Two are negative odd and three are positive even.

6. If |2 − x| = 3x − 6, then which of the following could be the value of x?

(A) -3   (B) -2   (C) 0   (D) 2   (E) 6

7. If x + 3 varies directly as y + 3 and x=9 when y=3, what is the value of x when y=6?

(A) 15   (B) 12   (C) 10   (D) 9   (E) 8

GO ON TO THE NEXT PAGE ▶▶▶

Model Test 7

8. In figure 1, if the length of SA is 10t-9 and SA=$\frac{1}{3}$AT, what is the length of ST

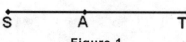

Figure 1
Figure not drawn to scale

(A) 27t – 30   (B) 30t – 27   (C) 36t – 40   (D) 40t – 36   (E) 46t – 30

9. Which of the following is not a solution of -5y + 3x > 15?

(A) (5, -1)   (B) (0,-4)   (C) (4, 0)   (D) (2, -3)   (E) (4, -2)

10. In figure 2, what is the value of s + t?

(A) 90   (B) 120   (C) 180   (D) 270   (E) 300

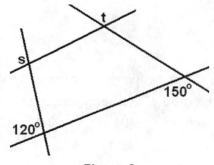

Figure 2
Figure not drawn to scale

11. A function f(x) is defined as f(x) = ax^2 + bx + c where a, b, and c are all real numbers. If f(x) and –f(x) intersect at no points then which of the following quantities must be nonzero?

   I. a   II. c   III. b^2 – 4ac

(A) I only   (B) I and II only   (C) I and III only   (D) II and III only   (E) I, II and III

12. If a is an arbitrarily chosen two digit prime number, which of the following quantities is not always odd?

(A) 2a-3   (B) 4a+7   (C) 10a+11   (D) (a+2)(a+4)   (E) 3a(3a+1)(3a+2)

GO ON TO THE NEXT PAGE ►►►

Model Test 7

13. If $f(x) = \sqrt{x+2}$ and $g(x) = x^2$, then $f(g(2.2)) =$

(A) 2.2   (B) 2.62   (C) 4.2   (D) 4.62   (E) 6.22

14. In rectangle ABCD in figure 3, what are the coordinates of vertex A?

(A) (1,3)   (B) (1,4)   (C) (2,3)   (D) (2,4)   (E) (4,1)

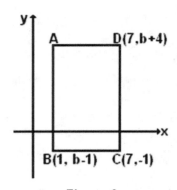

Figure 3
Figure not drawn to scale

15. If the ratio of the area of an equilateral triangle to the area of a regular hexagon is 2/3, then what is the ratio of the perimeter of the equilateral triangle to that of the regular hexagon?

(A) 4/9   (B) 2/3   (C) 1   (D) 3/2   (E) It cannot be determined by the given information

16. In figure 4 it is given that FE || GH || BA || CD; EH || BC and FG || AD. If the trapezoids ABCD and GHEF are similar, then a + b = ?

(A) -1   (B) -4.75   (C) -5.25   (D) -5.75

(E) It cannot be determined from the information given.

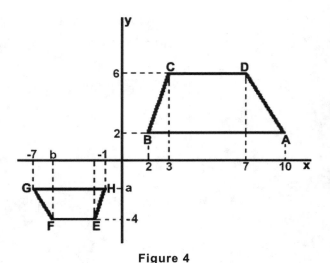

Figure 4
Figure not drawn to scale

GO ON TO THE NEXT PAGE ▶ ▶ ▶

Model Test 7

17. If an operation β is defined for all real numbers a and b by the equation

$$a \beta b = \frac{\text{least common multiple of a and b}}{\text{greatest common factor of a and b}} \text{ , then } 12 \beta 15 =$$

(A) 3   (B) 9   (C) 20   (D) 60   (E) 180

18. What is the sum of the integers that satisfy the inequality given by $x^2 < 81$?

(A) 63   (B) 54   (C) 45   (D) 36   (E) 0

19. In figure 5, 0 is the center of the circle with radius 10, and line BD is tangent to the circle at B. If the measure of $\angle$BAC is 30° and |BC|=|CD|, what is the length of BD?

(A) 10   (B) 14.14   (C) 17.32   (D) 24.49   (E) 30

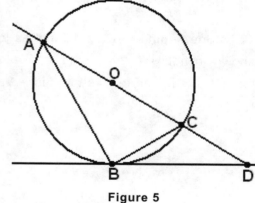

Figure 5

Figure not drawn to scale

20. 18 American, 12 German, and 10 Italian delegates have signed up for a conference. How many French delegates must sign up so that 20% of all delegates in the conference will be French?

(A) 8   (B) 10   (C) 12   (D) 14   (E) 16

21. If -16< m < 24 and 2 < n < 8, which of the following describes all possible values of m/n?

(A) (-2, 12)   (B) (3, 12)   (C) (-8, 3)   (D) (-8, 12)   (E) (-2, 3)

GO ON TO THE NEXT PAGE ▶▶▶

Model Test 7

22. In figure 6, ABCD and DEFC are isosceles trapezoids. If EF=2 and DC is 6, what is the length of AE given as x?

(A) 4   (B) 5   (C) 6   (D) 7

(E) The given information is not enough to find x.

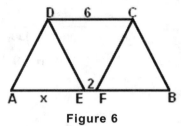

Figure 6

Figure not drawn to scale

23. If the average (arithmetic mean) of the set of real numbers {3, 8, 10, x, y} is 14.2, then x + y is

(A) 5.0   (B) 5.0   (C) 35.8   (D) 50   (E) 142

24. A line segment AB is 10 inches long and it is drawn within a plane P as shown in figure 7. All points at a distance of 5 inches from this segment in plane P are joined to give the figure Q. What is the perimeter of figure Q?

(A) 10   (B) 20   (C) 31.4   (D) 41.4   (E) 51.4

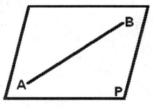

Figure 7

25. For the right triangle ΔABC given in figure 8, BC is 15 inches long and it is given that AB + AC = 25 inches. What is sin(x)?

(A) 8/15   (B) 17/8   (C) 8/17   (D) 17/15   (E) 15/17

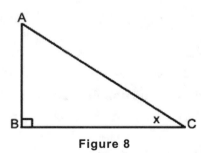

Figure 8

Figure not drawn to scale

26. If the cost of renting a car for x days is f(x), which of the following gives the cost of renting 5 cars for 3 days?

(A) 5f(x)   (B) 3f(x)   (C) 5f(3)   (D) f(15)   (E) 3f(5)

GO ON TO THE NEXT PAGE ▶▶▶

Model Test 7

Figure 9

27. On the line given in figure 9 above, if PQ > QR > RS > ST then which of the following may be false?

(A) PR > RS   (B) PR > QT   (C) PS > RT   (D) PS > ST   (E) QS > ST

28. Which of the lines in figure 10 must intersect in order to get the reflection of point (1,-2) about the y axis?

(A) n and p   (B) m and p   (C) m and s

(D) n and t   (E) m and r

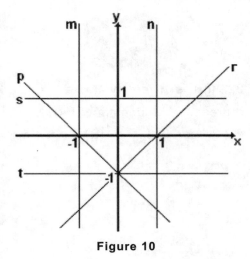

Figure 10

29. If all Acorageddons are Bourdeons and some Acorageddons are Clarimineans, then which of the following cannot be correct?

(A) Some Clarimineans are neither Bourdeons nor Acorageddons.

(B) Some Bourdeons are neither Acorageddons nor Clarimineans.

(C) Some Clarimineans are both Acorageddons and Bourdeons.

(D) Some Clarimineans are Bourdeons but not Acorageddons.

(E) Some Clarimineans are Acorageddons but not Bourdeons.

30. If ab + c = a, where a, b, and c are all distinct nonzero integers, which of the following must be true?

(A) a is a multiple of b   (B) a is multiple of b + c   (C) c is a multiple of a

(D) b is a multiple of c   (E) c is multiple of a − c

GO ON TO THE NEXT PAGE ▶▶▶

Model Test 7

31. In a discount store called "Cool Buy", only certain items are sold. Each item is labeled in such a way that the label starts with the letters "CB" followed by two capital letters and ends with a two digit number that represents one of the 14 different sections of the store. For example, CBRM07 is the label of a certain brand of shampoo whereas CBEA11 is the label of jasmine rice. How many distinct labels are possible?

(A) 676   (B) 6,760   (C) 9,464   (D) 67,600   (E) 45,697,600

32. f(x) is given by f(x) = $\sqrt{x-2}+2$ and the range of f(x) is 6 < f(x) < 11; which of the following represents the domain of f(x)?

(A) 2 < x < 3   (B) 4 < x < 5   (C) 18 < x < 83   (D) 38 < x < 123   (E) 66 < x < 171

33. In the figure 11, a regular pentagon has been divided into five congruent triangles. Which of the following statements is correct?

(A) |OA| < |CD| (B) |AC| > 2|ED| (C) Each triangle is equilateral.

(D) |AO|+|BO|+|CO|+|DO|+|EO| > the perimeter of ABCDE.

(E) Measure of angle OAB is greater the measure of angle AOB.

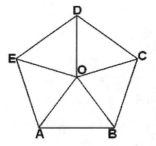

Figure 11

34. At a meeting, everyone shakes hands with every other person exactly once. If 66 handshakes take place, how many people are there at the meeting?

(A) 10   (B) 12   (C) 22   (D) 33   (E) 66

GO ON TO THE NEXT PAGE ▶ ▶ ▶

35. Points A, B, and C lie on a circle. If the center of that circle lies in the exterior of triangle ABC, then triangle ABC must be

(A) right   (B) acute   (C) obtuse   (D) scalene   (E) equilateral

36. If $i^2 = -1$, then $i^{35} + i^{36} + i^{37} =$

(A) -1 (B) i (C) $-i$ (D) 0   (E) 1

37. If a triangle ABC has sides of lengths 12, 11, and 8, then ABC must be

(A) an equilateral triangle   (B) a right scalene triangle   (C) an obtuse isosceles triangle

(D) an obtuse scalene triangle   (E) an acute scalene triangle

38. In figure 12, if the square has a side length of a, what is the ratio of the area of the circumscribed circle to that of the inscribed circle?

(A) $\frac{\sqrt{2}}{2}$ (B) $2\sqrt{2}$ (C) 2   (D) 0.5   (E) $\sqrt{2}$

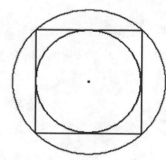

Figure 12

39. If the circumference of square ABCD is 4 times the circumference of square EFGH, what is the ratio of the area of square ABCD to that of square EFGH?

(A) 2:1   (B) 4:1   (C) 16:1   (D) 1:4   (E) 1:16

GO ON TO THE NEXT PAGE ▶▶▶

Model Test 7

40. For the functions $f(x) = x^2 + 5x - 2$ and $g(x) = 2x + 8$, what is the distance between the two points of intersection of their graphs?

(A) 16.56　　　　(B) 15.65　　　　(C) 12.12　　　　(D) 10.44　　　　(E) 9.54

41. The volume of the cube in figure 13 is 64. Which of the following is false?

(A) $5 < AC < 6$

(B) $6 < BH < 7$

(C) AHG is a right triangle.

(D) FHD is an isosceles triangle.

(E) BEG is an equilateral triangle.

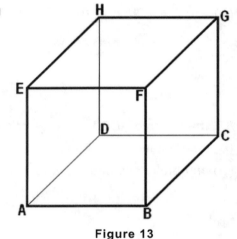

Figure 13

42. At the end of 2003, a certain car was worth $44,000. If the value of the car decreases at a rate of 3 percent each year, approximately how much will the car be worth at the end of 2012?

(A) $57,410　　　　(B) $34,484　　　　(C) $33,450　　　　(D) $32,446　　　　(E) $1,775

43. Which of the following sets of expressions correctly describes all points in the shaded region given in figure 14?

(A) $y = x^2$ and $y = 2 - x$ 　　　　(B) $y \geq x^2$ or $y < 2 - x$

(C) $y > x^2$ and $y \leq 2 - x$ 　　　　(D) $y > x^2$ or $y \leq 2 - x$

(E) $y \geq x^2$ and $y < 2 - x$

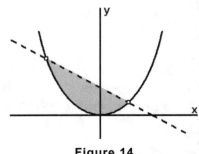

Figure 14

GO ON TO THE NEXT PAGE ▶ ▶ ▶

44. If the length of the shortest diagonal of a regular hexagon is t, what is the length of its longest diagonal?

(A) $t\sqrt{3}$ (B) $2t$ (C) $2t\sqrt{3}$ (D) $\dfrac{t}{\sqrt{3}}$ (E) $\dfrac{2t}{\sqrt{3}}$

45. If the value of $\log_5 x$ is between 2 and 3 then x cannot be

(A) 90 (B) 100 (C) 110 (D) 120 (E) 130

46. The first three terms of an arithmetic sequence are -4, 3, and 10. What is the 99th term of the sequence?

(A) 668 (B) 675 (C) 682 (D) 689 (E) 693

47. Triangular arrangements of numbers are shown in figure 15 and labeled as is given. Which of the following gives the sum of the numbers in the cells of the triangle for the n'th arrangement?

(A) $3n - 2$ (B) $\dfrac{n(n+1)}{2}$ (C) $\dfrac{n(n+1)(2n+1)}{6}$

(D) n^2 (E) $\dfrac{n^2(n^2+1)}{2}$

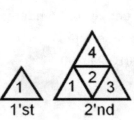
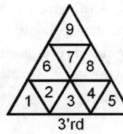

Figure 15

GO ON TO THE NEXT PAGE ▶▶▶

Model Test 7

For questions 48 through 50, please refer to the data given in the following graph:

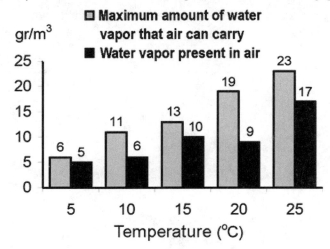

VI, vaporization index, is defined as the ratio of the maximum amount of water that air can carry to the water vapor present in air.

48. At which temperature, is VI closest to 2?

(A) 5   (B) 10   (C) 15   (D) 20   (E) 25

49. Based on the data given in the graph above, the median value of VI appears at which temperature?

(A) 5   (B) 10   (C) 15   (D) 20   (E) 25

50. At a certain temperature 11,793 grams of water vapor is present in 1,972 m^3 of air. What is the maximum amount, rounded to the nearest hundred grams, of water vapor that twice the given amount of air can carry at this temperature?

(A) 15,300   (B) 21,600   (C) 21,700   (D) 43,300   (E) 43,400

S T O P

END OF TEST

(Answers on page 201 – Solutions on page 217)

Model Test 8

Test Duration: 60 Minutes

Directions: For each of the following problems, decide which is the **best** of the choices given. If the exact numerical value is not one of the choices, select the choice that best approximates this value. Then fill in the corresponding oval on the answer sheet.

Notes:

- A calculator will be necessary for answering some (but not all) of the questions in this test. For each question you will have to decide whether or not you should use a calculator. The calculator you use must be at least a scientific calculator; programmable calculators and calculators that can display graphs are permitted.

- The only angle measure used on this test is degree measure. Make sure your calculator is in the degree mode.

- Figures that accompany problems in this test are intended to provide information useful in solving the problems. They are drawn as accurately as possible **except** when it is stated in a specific problem that its figure is not drawn to scale.

- All figures lie in a plane unless otherwise indicated.

- Unless otherwise specified, the domain of any function f is assumed to be the set of all real numbers **x** for which **f(x)** is a real number.

Reference Information: The following information is for your reference in answering some of the questions in this test.

- Volume of a right circular cone with radius **r** and height **h**: $V = \frac{1}{3}\pi r^2 h$

- Lateral area of a right circular cone with circumference of the base **c** and slant height **l**: $S = \frac{1}{2}cl$

- Volume of a sphere with radius **r**: $V = \frac{4}{3}\pi r^3$

- Surface area of sphere with radius **r**: $S = 4\pi r^2$

- Volume of a pyramid with base area **B** and height **h**: $V = \frac{1}{3}Bh$

1. If 4 times the reciprocal of p squared equals 1 then p can be

(A) -3   (B) -2   (C) -1   (D) 1   (E) 4

2. If $\frac{2x}{3} = \frac{3y}{2}$ then for what value of x does y equal x?

(A) 2/3   (B) 3/2   (C) 1   (D) 0   (E) 4/9

GO ON TO THE NEXT PAGE ▶▶▶

Model Test 8

3. If $\sqrt[5]{x} = 2.07$ then $\dfrac{160}{x} = ?$

(A) 38.01　　　　(B) 4.21　　　　(C) 0.23　　　　(D) 0.24　　　　(E) 0.026

4. The point A(-3, -7) is reflected across the x axis to get B. What are the coordinates of point B?

(A) (3, 7)　　　　(B) (3, -7)　　　　(C) (-3, 7)　　　　(D) (7, 3)　　　　(E) (7, -3)

5. If $41,000 = 4.1 \cdot 10^k$ then k =?

(A) 1　　　　(B) 6　　　　(C) -4　　　　(D) 4　　　　(E) -6

6. $\left(\sqrt{7} - \sqrt{3}\right) \cdot \left(\sqrt{3} + \sqrt{7}\right) = ?$

(A) -7　　　　(B) -4　　　　(C) -3　　　　(D) 3　　　　(E) 4

7. If $a = 2b = 3c = 4$ then $2a^2 - 5 + 3b - 4c = ?$

(A) 26.6　　　　(B) 26.7　　　　(C) 27.6　　　　(D) 27.7　　　　(E) 27.8

8. If 17 bourons make 2 acorageddons and 5 acorageddons are sold for 137 cents then what will be the cost in cents of the acorageddons made from p bourons?

(A) $\dfrac{85}{274p}$　　　　(B) $\dfrac{85p}{274}$　　　　(C) $\dfrac{274p}{85}$　　　　(D) $\dfrac{274}{85p}$

(E) None of the above

GO ON TO THE NEXT PAGE ▶▶▶

Model Test 8

9. If $x^2 + 4x + 4y = y^2$ and y = -2 then x can be

(A) -6   (B) -2   (C) 3   (D) 4   (E) 6

10. The function f(x) is given by $f(x) = \dfrac{x^2}{3} - \ln(2x)$. Which of the following is a zero of f(x)?

(A) 0.54   (B) 0.56   (C) 2.06   (D) 2.09   (E) 2.60

11. What does the equation given by $2x^2 + 2y^2 + 12x - 16y + 50 = 0$ represent in the cartesian plane?

(A) A point   (B) A circle   (C) An ellipse   (D) A hyperbola   (E) A parabola

12. In figure 1, if $\cos\theta = \dfrac{3}{5}$ for the given right triangle, what is the area of the triangle?

(A) 6   (B) 12   (C) 24   (D) 48   (E) 60

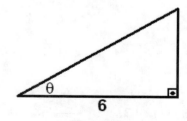

Figure 1

Figure not drawn to scale

13. If f(x) is a linear function with a slope of 5 and f(2) = 17, then f(0) =

(A) 2   (B) 5   (C) 7   (D) 10   (E) 15

14. $x^2 - 6x + 9 = (x + 1)^2 + b(x + 1) + c$ for all values of x, then c =

(A) -8   (B) -5   (C) 2   (D) 7   (E) 16

GO ON TO THE NEXT PAGE ▶▶▶

15. If g(x) = 2x − 1 and f(g(x)) = 3x + 4 then f(x)=?

(A) $\dfrac{3x+5}{2}$ (B) $\dfrac{2x-5}{3}$ (C) $\dfrac{3x+11}{2}$ (D) $\dfrac{2x-11}{3}$

(E) None of the above

16. If y = $\dfrac{3-x}{9-x^2}$ then for what values of x is y not defined?

(A) for all real numbers except for 3   (B) for 3 only

(C) for all real numbers except for -3   (D) for -3 and 3 only

(E) for all real numbers except for -3 and 3

17. What are the x intercepts of the function given by y = x³ − 5x² − 6x?

(A) 0   (B) 0, 6   (C) 0, 6, -1   (D) 0, 2, 3   (E) 0, -2, -3

18. $\sin x \cdot \left(3\sin x + \dfrac{2}{\sin x} \right) + 3\cos^2 x = ?$

(A) 2   (B) 3   (C) 4   (D) 5   (E) undefined

19. If α is an acute angle for which $\cos\alpha = \dfrac{3}{y}$ then tanα=?

(A) $\dfrac{\sqrt{y^2-9}}{3}$ (B) $\dfrac{3}{\sqrt{y^2-9}}$ (C) $\dfrac{\sqrt{y^2-9}}{y}$ (D) $\dfrac{y^2-9}{9}$ (E) $\dfrac{y}{3}$

GO ON TO THE NEXT PAGE ▶▶▶

Model Test 8

20. A regular hexagon is given in figure 2 and each side of the shaded triangle measures 12 inches. What is the area of the regular hexagon in square feet?

(A) $36\sqrt{3}$ (B) $72\sqrt{3}$ (C) $\dfrac{\sqrt{3}}{4}$ (D) $\dfrac{\sqrt{3}}{2}$ (E) $\sqrt{3}$

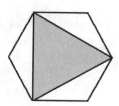

Figure 2

21. Which of the following circles allows the point (3, 2) to be outside it?

(A) $(x - 1)^2 + (y - 1)^2 = 9$ (B) $x^2 + (y - 1)^2 = 9$ (C) $x^2 + y^2 = 16$

(D) $(x + 1)^2 + (y + 1)^2 = 25$ (E) $x^2 + (y + 1)^2 = 36$

22. If A, B and C are three of the eight vertices of the cube given in figure 3 then what is the degree measure of the marked angle ABC?

(A) 35.3°   (B) 37.7°   (C) 38.7°   (D) 42.3°   (E) 44.5°

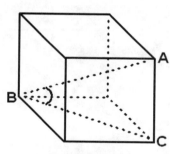

Figure 3

23. If ABC in figure 4 is a right triangle where B is the right angle and AC = x + 4 then what is the perimeter of this triangle?

(A) 24   (B) 26   (C) 28   (D) 30

(E) It cannot be determined from the information given.

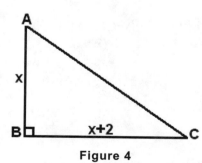

Figure 4

Figure not drawn to scale

GO ON TO THE NEXT PAGE ▶▶▶

Model Test 8

24. If a + b = 2 and b + c = 4 then $\dfrac{b+c}{a+3b+2c} = ?$

(A) 0.25   (B) 0.4   (C) 2.5   (D) 4   (E) 40

25. The graph of f(x) is given in figure 5. Which of the following is the graph of f(x+1)?

(A)

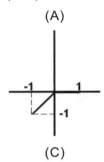

(B)

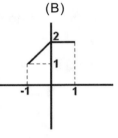

Figure 5

(C)

(D)

(E)

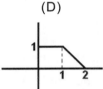

26. In figure 6 O is the center of the circle and minor arcs AB and BC measure 90° and 150° respectively. The arrow that freely rotates around O is designed in such a way that it can never stop on a border line. Erol will spin the arrow twice; what is the probability that it will land on region 2 at least once?

(A) 1/4   (B) 3/4   (C) 7/16   (D) 9/16   (E) 15/16

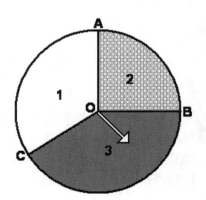

Figure 6
Figure not drawn to scale

GO ON TO THE NEXT PAGE ▶ ▶ ▶

27. A function f(x) is given by f(x) = x^2 + 3x + k where k is a randomly selected integer from the set {1, 2, 3, 4, 5, 6, 7, 8}. What is the probability f(x) intersects the x axis at one or more points?

(A) 1/8   (B) 1/4   (C) 3/8   (D) 5/8   (E) 3/4

28. If lna = 2, logb = -1 and $\log_2 c = 3$ then abc = ?

(A) $\dfrac{5}{4e^2}$ (B) $\dfrac{4e^2}{5}$ (C) $\dfrac{4}{5e^2}$ (D) $20e^2$ (E) $80e^2$

29. If it takes one third of a minute to produce n articles on a production line, then in terms of n, how many articles can be produced at in half an hour at twice this rate?

(A) 45n   (B) 90n   (C) 135n   (D) 180n   (E) 225n

30. How many different ordered pairs (x, y) with positive components can each be a solution to the equation given by $18+x+2y^2=4x-4y^2$?

(A) None   (B) One   (C) Two   (D) Four   (E) More than four

31. What is the minimum value of the sum of a number and twice its reciprocal given that the number is positive?

(A) 1.04   (B) 1.41   (C) 2.08   (D) 2.82   (E) 2.83

GO ON TO THE NEXT PAGE ▶▶▶

Model Test 8

32. If lines m and n are parallel in figure 7 then which of the following cannot be deduced?

(A) $\hat{1} = \hat{5}$ (B) $\hat{2} = \hat{8}$ (C) $\hat{1} + \hat{3} = \hat{8}$ (D) $\hat{1} + \hat{6} = 180°$ (E) $\hat{1} + \hat{2} = \hat{4} + \hat{5}$

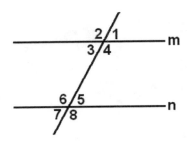

Figure 7
Figure not drawn to scale

33. If $y^6 \cdot z^5 = (yz)^5$ and $y \neq z$ then y can be

(A) 0 only

(B) 1 only

(C) 0 or 1 only

(D) any nonzero real number

(E) any nonzero real number other than 1

34. For the rectangle ABCD given in figure 8 9AE=4EF=6FB=36 and AD=10. If the rectangle will be folded along the dotted lines m and n so that regions 1 and 3 will partially overlap; then what will be the area of the overlap?

(A) 10   (B) 20   (C) 30   (D) 40   (E) 60

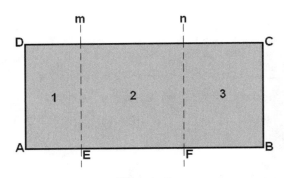

Figure 8
Figure not drawn to scale

35. A number N is defined as N = $a^2 \cdot b^3$ where a and b are different prime numbers greater than 2. How many distinct integer divisors does N have?

(A) 5   (B) 10   (C) 12   (D) 18   (E) 24

GO ON TO THE NEXT PAGE ▶▶▶

Model Test 8

36. If the 5th and the 11th term of a geometric sequence are 64 and 1 respectively then what is the 3rd term?

(A) 128　　　　　(B) 256　　　　　(C) 512　　　　　(D) 1024　　　　　(E) 2048

37. If each side of the cube given in figure 9 measures 18 inches what is the perimeter in feet of the shaded triangle?

(A) 5.4　　　(B) 5.5　　　(C) 43.9　　　(D) 65.9　　　(E) 66.0

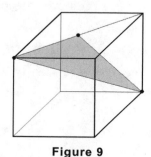

Figure 9

Figure not drawn to scale

38. The circular wave front given in figure 10 expands in such a way that its radius in meters increases as a function of time by the formula $R(t) = 4.5t$ where t is the number of seconds that has passed since the wave front was first created by a stone that fell into the water. If n is the integer number of seconds after which the perimeter of the wave front exceeds 100 meters for the first time then n = ?

(A) 3　　　　(B) 4　　　　(C) 6　　　　(D) 8　　　　(E) 12

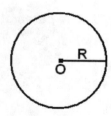

Figure 10

39. If a line contains the point (-2, 1) but does not contain (-1, 2), the x – intercept of this line can not be

(A) -3　　　　　(B) -1　　　　　(C) 0　　　　　(D) 1　　　　　(E) 3

40. If $3 < x < 7$ and $-2 < y < 8$, then

(A) $0 < xy < 56$　　(B) $-14 < xy < 56$　　(C) $-6 < xy < 56$　　(D) $1 < xy < 15$　　(E) $-14 < xy < 15$

GO ON TO THE NEXT PAGE ▶▶▶

Model Test 8

41. What is the perimeter of the square defined by the relation $|x| + |y| = 4$?

(A) 16   (B) 32   (C) 64   (D) $16\sqrt{2}$ (E) $32\sqrt{2}$

42. If the radius of the circle given in figure 11 is 5, what is the equation of the circle?

(A) $(x - 5)^2 + (y - 5)^2 = 5$ (B) $(x + 5)^2 + (y + 5)^2 = 5$

(C) $(x - 5)^2 + (y - 5)^2 = 25$ (D) $(x + 5)^2 + (y + 5)^2 = 25$

(E) $(x + 5)^2 + (y + 5)^2 + 25 = 0$

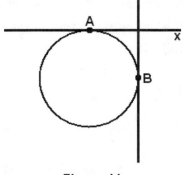

Figure 11

43. The parametric curve defined by $x = t^2$ and $y = -t^2 + 1$ represents

(A) a circle   (B) portion of a line   (C) a line   (D) a parabola   (E) portion of a parabola

44. If $3 + 2i$ is one root of a quadratic equation given by $x^2 - Px + Q = 0$ where P and Q are real numbers then Q is

(A) 0   (B) 6   (C) 13   (D) – 6   (E) – 13

45. How many quarts of pure water must be added to 60 quarts of a 15% acid solution in order to obtain a 12% acid solution?

(A) 13   (B) 14   (C) 15   (D) 18   (E) 30

GO ON TO THE NEXT PAGE ▶▶▶

Model Test 8

46. A theorem states that if parallel segments are drawn to each of the three sides of an equilateral triangle ABC from any point P inside the triangle, as in figure 12, then the sum of the lengths of the parallel segments will be a constant. This constant is equal to

(A) the altitude of ABC

(B) the perimeter of ABC

(C) half the perimeter of ABC

(D) the length of a side of ABC

(E) $\sqrt{3}$ times the length of a side of ABC

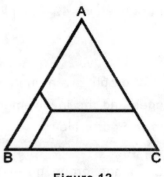

Figure 12

47. Based on the information given in figure 13 x = ?

(A) 13.4   (B) 13.5   (C) 13.6   (D) 13.7

(E) It cannot be determined from the information given.

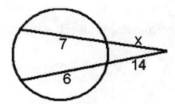

Figure 13

Figure not drawn to scale

48. The front view of **RUSH** Building is given in figure 14. Two trees each having a height of h=7 meters are 99 meters apart. If the height of the building is H=77 meters and the angles of elevation α and β in degrees are 55° and 66° respectively, then what is the width of the building in meters?

(A) 10.8   (B) 13.8   (C) 18.8

(D) 19.8   (E) 28.8

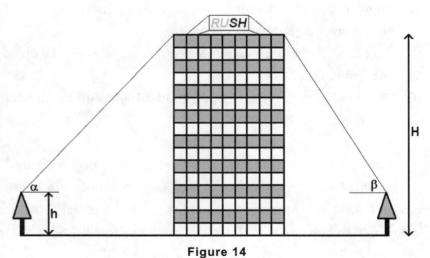

Figure 14

Figure not drawn to scale

GO ON TO THE NEXT PAGE ▶▶▶

Model Test 8

For questions 49 – 50 please refer to the following graph.

The graph given in figure 15 represents the ages and the involvement in sports activities in town Sportsville.

Each age group given to be x represents all ages in the interval x − 5 ≤ age ≤ x + 4.

For example there are 5000 people in the age group 30 that represents all ages in the interval 25 ≤ age ≤ 34 and 4000 people in this age group are involved in sports activities.

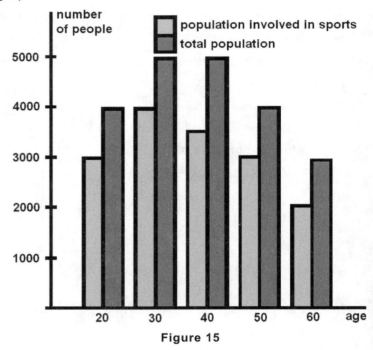

Figure 15

49. Based on the data given in the graph in figure 15 above, which of the following is a false statement?

(A) About 21000 people in town Sportsville are at least 15 and at most 64 years old.

(B) About 15500 people in town Sportsville are involved in sports activities.

(C) About 73.8 percent of the people in town Sportsville who are at least 15 and at most 64 years old are involved in sports activities.

(D) Percentage of the people in age group 30 who are involved in sports activities is greater that in the age group 20.

(E) The number of people in age group 20 who are involved in sports activities is less that in the age group 40.

50. A recent immigration has increased the age group 30 by 1500 people; however the percentage of people involved in sports among the new comers is the same as the previous percentage in this age group. If 15% of the new comers involved in sports activities are the only people who prefer to play golf in town, then how many golf players are there totally?

(A) 140   (B) 160   (C) 180   (D) 200   (E) 220

S T O P

END OF TEST

(Answers on page 201 – Solutions on page 219)

Model Test 9

Test Duration: 60 Minutes

Directions: For each of the following problems, decide which is the **best** of the choices given. If the exact numerical value is not one of the choices, select the choice that best approximates this value. Then fill in the corresponding oval on the answer sheet.

Notes:

- A calculator will be necessary for answering some (but not all) of the questions in this test. For each question you will have to decide whether or not you should use a calculator. The calculator you use must be at least a scientific calculator; programmable calculators and calculators that can display graphs are permitted.

- The only angle measure used on this test is degree measure. Make sure your calculator is in the degree mode.

- Figures that accompany problems in this test are intended to provide information useful in solving the problems. They are drawn as accurately as possible **except** when it is stated in a specific problem that its figure is not drawn to scale.

- All figures lie in a plane unless otherwise indicated.

- Unless otherwise specified, the domain of any function f is assumed to be the set of all real numbers **x** for which **f(x)** is a real number.

Reference Information: The following information is for your reference in answering some of the questions in this test.

- Volume of a right circular cone with radius **r** and height **h**: $V = \frac{1}{3}\pi r^2 h$

- Lateral area of a right circular cone with circumference of the base **c** and slant height **l**: $S = \frac{1}{2}cl$

- Volume of a sphere with radius **r**: $V = \frac{4}{3}\pi r^3$

- Surface area of sphere with radius **r**: $S = 4\pi r^2$

- Volume of a pyramid with base area **B** and height **h**: $V = \frac{1}{3}Bh$

1. If $x = \sqrt{7}$, then $\sqrt{56 - x^2} =$

(A) x   (B) 2x   (C) x^2 (D) $x + \sqrt{7}$ (E) x+7

2. In figure 1, the area of the shaded region is what percent of the area of square ABCD?

(A) 12.5%   (B) 37.5%   (C) 20%   (D) 40%   (E) 25%

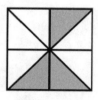

Figure 1

GO ON TO THE NEXT PAGE ▶▶▶

Model Test 9

$$x - y^2 = 6$$
$$3x - y = 42$$

3. If (x,y) is the solution to the above system of equations, then what is one possible value of x?

(A) -3   (B) 3   (C) -15   (D) 15   (E) No Solution

4. The cost **c** of operating a production line with respect to the number **x** of the items produced, is a function given by $c = 937 + 19.9x$. How many items will have been produced when the cost of the production is $2529?

(A) 80   (B) 120   (C) 160   (D) 40

(E) None of the above

5. If $xy+y=15$ and $x^2+2=11$, what is one possible value of y?

(A) 5   (B) $\dfrac{15}{4}$ (C) 4   (D) $\dfrac{15}{2}$ (E) 3

6. The graph given in figure 2 is the set of all x such that

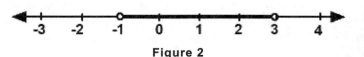

Figure 2

(A) $|x-1| < 2$ (B) R \ (-1,3)

(C) $|x-1| > 2$ (D) R \ [-1,3]

(E) x> -1 or x<3

7. If a represents an odd integer, which one(s) of the following represent(s) even integers?

    I. a^2-2a

    II. $(a+2)(a-2)$

    III. a^3-3a

(A) I only   (B) II only   (C) III only   (D) I and II only   (E) II and III only

GO ON TO THE NEXT PAGE ▶▶▶

8. In figure 3, if ABCD is a square with a side length of 3, the coordinates of point P are

(A) (0,6)   (B) (6,0)   (C) (0,-6)   (D) (-6,0)

(E) It cannot be determined from the information given.

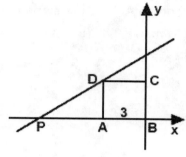

Figure 3

9. What are all values of x for which $x < x^3 < x^2$?

(A) $x > 1$ (B) $x \geq 1$ (C) $0 < x < 1$ (D) $-1 \leq x < 1$ (E) $-1 < x < 0$

10. If $x \neq 0$, then $\dfrac{x+x+x}{x \cdot x} =$

(A) x (B) 3/2   (C) 3/x   (D) $x^2/2$ (E) x^2

11. The polygon given in figure 4 can be divided into three congruent equilateral triangles, each with side length t inches. If the perimeter of the polygon is 28 inches, what is the value of t?

(A) 3   (B) 3.11   (C) 4   (D) 4.64   (E) 9.33

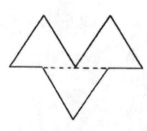

Figure 4

12. Hilmi initially has $70. He earns $5 per hour before 2:00 PM and $7 per hour after 2:00 PM. Which of the following represents Hilmi's total money after working between 08:00 AM to 5:00 PM in one day?

(A) $5 · 8 + $7 · 5   (B) $5 · 6 + $7 · 3   (C) $70 + $5 · 8 + $7 · 5

(D) $70 + $5 · 6 + $7 · 3   (E) $80 + $5 · 6 + $7 · 3

GO ON TO THE NEXT PAGE ▶▶▶

Model Test 9

13. When triangle ABC in figure 5 is reflected across the x axis to get triangle A'B'C', the slope m of the side A'C' will satisfy the relation

(A) m<0   (B) m=0   (C) 0<m<1   (D) m=1   (E) m>1

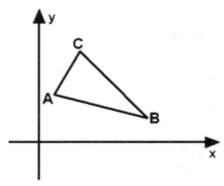

Figure 5

14. Which of the values a, b, c, and d, is sufficient by itself in order to determine the area of the square given in figure 6?

(A) a   (B) b   (C) c   (D) d

(E) None of the above

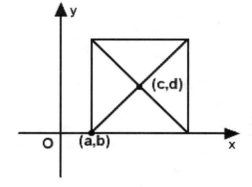

Figure 6

15. Which of the following points are on the line with the equation $y = \frac{1}{2}x - 2$?

    I. $\left(1, \frac{3}{2}\right)$ II. $(2, -1)$ III. $(6, 1)$

(A) I only   (B) I and III only   (C) II and III only

(D) III only   (E) I, II and III

16. If x=1-a and x-y=2, then xy=

(A) -1-a   (B) a-1   (C) 1-a²   (D) a²-1   (E) 1+a²

GO ON TO THE NEXT PAGE ▶▶▶

17. For what value of x does the equation $5^{x-2} = \dfrac{1}{125}$ hold?

(A) 1/5   (B) 0.5   (C) 0   (D) -1   (E) -2

18. In a classroom, there are n students totally and, g of them are girls. If 15% of the boys in the classroom are at 17 years of age or older, how many boys under 17 years of age are there in this classroom?

(A) 0.85(n-g)   (B) 0.83.(n-g)   (C) 0.85.(g-n)   (D) $\dfrac{0.85(n-g)}{2}$ (E) $\dfrac{0.85(n-g)}{n}$

19. If $f(x)=3x^2-ax+5$ and $f(1)=4$, then $f(-1)=$

(A) -4   (B) 0   (C) 4   (D) 8   (E) 12

20. A bug moves along the x axis starting at -3 and moves steadily leftward to -5 and then steadily rightward to +13. What is the total distance the bug traveled?

(A) 18   (B) 20   (C) 16   (D) 44   (E) 2

21. In figure 7, B is the midpoint of segment AC, and C is $\dfrac{3}{5}$ of the way from B to D. What is the length of segment AB in inches if AD is 16 inches long?

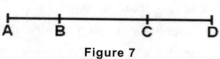

Figure 7
Figure not drawn to scale

(A) 2   (B) 4   (C) 6   (D)10   (E)12

GO ON TO THE NEXT PAGE ▶▶▶

Model Test 9

$$y_1 = (3-x)(x-7)$$

$$y_2 = -x \cdot (4+x)$$

$$y_3 = (5-x) \cdot (10-5x)$$

$$y_4 = \frac{2-x}{x+3}$$

$$y_5 = 2^{5-x}$$

22. How many of the above functions have a negative y intercept?

(A) 1   (B) 2   (C) 3   (D) 4   (E) 5

23. In figure 8, AB is a diameter of the circle. Then |AB| is

(A) 4   (B) 5   (C) 8   (D)10

(E) It cannot be determined from the information given

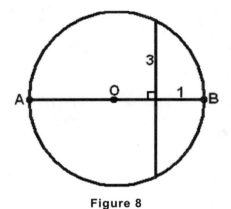

Figure 8

Figure not drawn to scale

24. Let S be the supplement and C be the complement of an acute angle A. The difference between S and C

(A) equals 90°.

(B) is equal to the measure of A.

(C) is less than the measure of A.

(D) equals twice the measure of A.

(E) must be greater than twice the measure of A.

GO ON TO THE NEXT PAGE ▶▶▶

Model Test 9

For questions 25 – 26 please refer to the data given in the following graph:

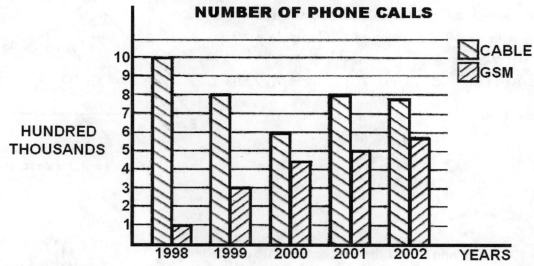

NUMBER OF PHONE CALLS

25. When is the difference between the number of calls via cable and GSM the least?

(A) 1998   (B) 1999   (C) 2000   (D) 2001   (E) 2002

26. When is the total number of calls, the greatest?

(A) 1998   (B) 1999   (C) 2000   (D) 2001   (E) 2002

27. The lengths of the sides of a triangle are 7, 24 and 25. If θ is the measure of the greatest acute angle in the triangle, what is the value of $\sin\theta$?

(A) 7/24   (B) 24/25   (C) 24/7   (D) 25/24   (E) 25/7

28. For which of the following lists of numbers are the mean, median, and mode equal?

(A) -3, 1, 1, 2   (B) 1, 2, 2, 4   (C) 2, 3, 4, 6   (D) 3, 4, 4, 5   (E) 3, 4, 5, 7

GO ON TO THE NEXT PAGE ▶▶▶

Model Test 9

29. In figure 9, ∆ABC has lengths in inches and angle measures in degrees as shown. In ∆KLM, not shown, the measure of ∠K is 55.8°, the measure of ∠L is 41.4°, and KM is 12 inches. What is length in inches of LM?

   (A) 11.97   (B) 14.98   (C) 15.00   (D) 18.02   (E) 20.93

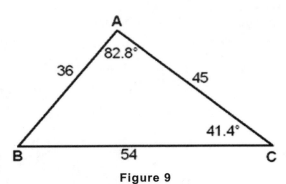

Figure 9
Figure not drawn to scale

30. In figure 10, ABCD and KLMN are two congruent squares each having a side length of a. What is the length of |PR| in terms of a?

   (A) $\dfrac{a}{2}$

   (B) $\dfrac{a}{\sqrt{2}} - \dfrac{a}{\sqrt{3}}$

   (C) $\dfrac{a}{\sqrt{2}} - \dfrac{a}{2}$

   (D) $\dfrac{a}{\sqrt{2}}$

   (E) $\dfrac{a}{\sqrt{3}} - \dfrac{a}{2}$

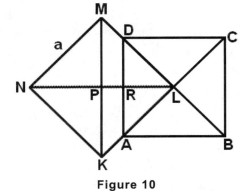

Figure 10

31. A car is sold for $50.000 and has an estimated life of 15 years, at the end of which its price is expected to be $5000. If the relation between the price of the car and the number of the years since its production is linear what will be the price of the car at the end of 5 years?

   (A) $25000   (B) $30000   (C) $35000   (D) $12500   (E) $15000

32. If (ax+b)(cx+d)=5x²-10x+7 for all values of x, what is the value of ac+bd?

   (A) -5   (B) -3   (C) 0   (D) 5   (E) 12

GO ON TO THE NEXT PAGE ▶▶▶

Model Test 9

33. A light in a machine blinks every 7 minutes. If it first blinks at 11:12AM, at what time does it blink for the 22nd time?

(A) 1:25AM   (B)1:39AM   (C) 1:46AM   (D) 1:39PM   (E) 1:46PM

34. The graph of which of the following lines has both an x-intercept of -3 and a y-intercept of -3?

(A) y=-3x-3   (B) y= -x-3   (C) y=3x+3   (D) y=x+3   (E) y=x-3

35. If f(x) is a quadratic function and f(-5) > f(-1), then which of the following must be correct?

(A) f(1) < f(2)   (B) f(3) > f(0)   (C) f(-2) < f(2)   (D) f(-3) > f(-2)   (E) f(-3) > f(2)

36. Given the graph of the circle whose equation is $(x-2)^2+(y+1)^2=4$, which of the following is an endpoint of the diameter with slope 1?

(A) (4, 1)   (B) (4, -1)   (C) (2, -1)   (D) $(2+\sqrt{2}, -1+\sqrt{2})$ (E) $(2-\sqrt{2}, -1+\sqrt{2})$

37. A cube with a surface area 54 square centimeters must fit through a circular opening whose radius is r centimeters. What is the smallest possible value of r in cm?

(A) 4.24   (B) 2.12   (C) 6.36   (D) 3.18   (E) 2.18

38. If f(x)=$\dfrac{x-1}{2}$ and f(g(2x))=2x, then g(3)=

(A) 1   (B) 7   (C) 2   (D) 0   (E)-1

GO ON TO THE NEXT PAGE ▶▶▶

Model Test 9

39. Figure shows a ladder whose bottom is 6 feet away from a vertical wall. If the angle of elevation of the ladder is 53°, what is the length of the ladder?

(A) 3 feet   (B) 4 feet   (C) 8 feet   (D) 10 feet   (E) 12 feet

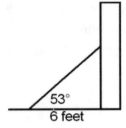

Figure 11
Figure not drawn to scale

40. x and y are positive in the system of equations $(xy)^{\frac{3}{4}}=2y^{\frac{1}{4}}$ and $(xy)^{\frac{1}{4}}=1.2x^{\frac{1}{4}}$. What is the value of x?

(A) 1.54   (B) 1.55   (C) 2.0   (D) 2.1   (E) 20.7

41. $14\sin^2x+(2\cos x)\cdot(7\cos x)=$

(A) 14   (B) 28   (C) 28sinxcosx   (D) $14\sin^2x+14\cos x$ (E) $14\sin^2x+14\cos^2(x^2)$

42. In figure 12, three parallel lines are intersected by two other lines. What is the value of x?

(A) 0.8   (B) 3   (C) 4   (D) 2   (E) 1

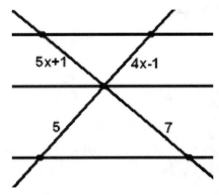

Figure 12
Figure not drawn to scale

43. It is given that x varies inversely as the square of y and x=4 when y=3. What is the value of x when y=2?

(A) 2   (B) 3   (C)6   (D) 9   (E) 11

GO ON TO THE NEXT PAGE ▶▶▶

Model Test 9

44. A line m and three distinct points A, B and C lie in the same plane. If m does not intersect segment AB and intersects segment BC at one and only one point, which of the following must be correct?

(A) m is perpendicular to line BC

(B) m is parallel to line AB

(C) m is not parallel to line AC

(D) A, B and C are not collinear

(E) A, B and C are collinear

45. If f is the function given by $f(x)=\dfrac{1}{\sqrt{3-x^2}}$, then the domain of f consists of all real numbers x such that

(A) $x \geq \sqrt{3}$ (B) $0 < x \leq \sqrt{3}$ (C) $0 \leq x < \sqrt{3}$ (D) $-\sqrt{3} < x < \sqrt{3}$ (E) $-\sqrt{3} \leq x \leq \sqrt{3}$

46. Two distinct line segments that entirely lie within the shaded region given in figure 13 can partition it into n non-overlapping sections; n can be

   I. 3

   II. 4

   III. 5

(A) I only

(B) II only

(C) I and II only

(D) II and III only

(E) I, II and III

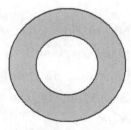

Figure 13

47. A number is called "unitary" when its digits are added, when the digits of the resulting number are added, and so on, the final result is 1. For example 7768 is a unitary number because, 7+7+6+8=28; 2+8=10; and finally 1+0=1. A sequence of unitary numbers starts with 775 and each number in the sequence is the next greater unitary number. What is the 57th term in this sequence?

(A) 1269   (B) 1270   (C) 1279   (D) 1288   (E) 2197

GO ON TO THE NEXT PAGE ▶▶▶

Model Test 9

48. By taking cross-sections from a right circular cylinder with one plane, how many of the following can be obtained?

    I. a circle   II. a line segment   III. an ellipse

    IV. a point   V. a rectangle   VI. a parabola

(A) 2   (B) 3   (C) 4   (D) 5   (E) 6

49. A spherical balloon is inflated in such a way that its radius increases at a constant rate of 24 inches per second. By how much, in feet, does the circumference of the largest cross-section increase from the beginning of the 3rd second to the end of the 5th second?

(A) 144π (B) 12π (C) 8π (D) 96π

(E) None of the above

50. In figure 14, ABCD and BHGK are squares with sides of length 4 and 9 respectively. Segment AG intersects segment BH at point E and EF is parallel to BK. What is the sum of the areas of the shaded regions?

(A) 28.5   (B) 38.5   (C) 48.5   (D) 58.5

(E) It cannot be determined from the information given.

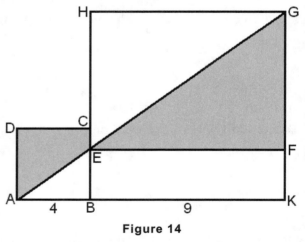

Figure 14

Figure not drawn to scale

S T O P

END OF TEST

(Answers on page 202 – Solutions on page 221)

Model Test 10

Test Duration: 60 Minutes

Directions: For each of the following problems, decide which is the **best** of the choices given. If the exact numerical value is not one of the choices, select the choice that best approximates this value. Then fill in the corresponding oval on the answer sheet.

Notes:

- A calculator will be necessary for answering some (but not all) of the questions in this test. For each question you will have to decide whether or not you should use a calculator. The calculator you use must be at least a scientific calculator; programmable calculators and calculators that can display graphs are permitted.

- The only angle measure used on this test is degree measure. Make sure your calculator is in the degree mode.

- Figures that accompany problems in this test are intended to provide information useful in solving the problems. They are drawn as accurately as possible **except** when it is stated in a specific problem that its figure is not drawn to scale.

- All figures lie in a plane unless otherwise indicated.

- Unless otherwise specified, the domain of any function f is assumed to be the set of all real numbers **x** for which **f(x)** is a real number.

Reference Information: The following information is for your reference in answering some of the questions in this test.

- Volume of a right circular cone with radius **r** and height **h**: $V = \frac{1}{3}\pi r^2 h$

- Lateral area of a right circular cone with circumference of the base **c** and slant height **l**: $S = \frac{1}{2}cl$

- Volume of a sphere with radius **r**: $V = \frac{4}{3}\pi r^3$

- Surface area of sphere with radius **r**: $S = 4\pi r^2$

- Volume of a pyramid with base area **B** and height **h**: $V = \frac{1}{3}Bh$

1. If $\left(\sqrt{x} \cdot \sqrt[3]{y}\right)^6 = 1728$ and x and y are positive integers then x+y can be

(A) 10   (B) 11   (C) 12   (D) 13   (E) 14

2. If $3a = b$ and $c = \frac{b}{4}$ then $\frac{a}{c} = ?$

(A) $\frac{3}{4}$ (B) $\frac{4}{3}$ (C) $\frac{1}{12}$ (D) 12

(E) None of the above

GO ON TO THE NEXT PAGE ▶▶▶

Model Test 10

3. If $f(A) = A^{3^2}$ then f(2.1) = ?

(A) 794.2   (B) 794.3   (C) 85.7   (D) 85.8   (E) 40.8

4. An object is projected vertically upwards so that t seconds later its height in meters is given by $h(t) = 100 + 20t - 5t^2$. The initial height of the object is

(A) 40 meters   (B) 75 meters   (C) 100 meters   (D) 115 meters   (E) 120 meters

5. The point A(-3, -7) is reflected across the y axis to get B. What are the coordinates of point B?

(A) (3, 7)   (B) (3, -7)   (C) (-3, 7)   (D) (7, 3)   (E) (7, -3)

6. If $0.0000034 = 3.4 \cdot 10^k$ then k =?

(A) 4   (B) 5   (C) -4   (D) -5   (E) -6

7. The seven digit integer given by 2,165,E78 is divisible by 9. E can be

(A) 4   (B) 5   (C) 7   (D) 8   (E) 9

8. A number N is defined as $N = a^3 \cdot b^4$ where a and b are different prime numbers greater than 2. How many distinct positive integer divisors does N have?

(A) 7   (B) 9   (C) 12   (D) 18   (E) 20

GO ON TO THE NEXT PAGE ▶▶▶

Model Test 10

9. For the right triangle given in figure 1, $\sin\alpha$ = ?

(A) $\cos\alpha$ (B) $\tan\alpha$ (C) $\sin\beta$ (D) $\cos\beta$ (E) $\tan\beta$

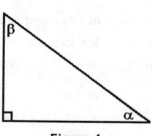

Figure 1

10. If p and q are irrational numbers, which of the following must also be irrational?

(A) $|p| + |q|$ (B) $\dfrac{p}{q}$ (C) pq (D) $\dfrac{p}{\pi} + \dfrac{q}{e}$ (E) $p\sqrt{2} - q\sqrt{3}$

11. If $x = \dfrac{-2y}{y-z}$ then what is y in terms of x and z?

(A) $y = -\dfrac{2x}{x-z}$ (B) $y = -\dfrac{x \cdot z}{x+2}$ (C) $y = \dfrac{x \cdot z}{z+2}$ (D) $y = \dfrac{x \cdot z}{x+2}$ (E) $y = \dfrac{2x}{x-z}$

12. $\dfrac{1}{k(k+1)} = \dfrac{A}{k} - \dfrac{B}{k+1}$ for every positive integer k; A + 3B=?

(A) 1   (B) 2   (C) -2   (D) -4   (E) 4

13. What is the slope of line AB given in figure 2?

(A) 0.36   (B) 0.83   (C) 0.84   (D) 2.74   (E) 2.75

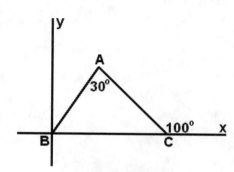

Figure 2

Figure not drawn to scale

GO ON TO THE NEXT PAGE ▶▶▶

Model Test 10

14. Line m passes through the origin and is perpendicular to the line given by $y + 2x = b$. If $(r, r-1)$ is a point on line m then $r = ?$

(A) 1   (B) 2   (C) 3   (D) 4   (E) 5

15. How many real solutions does the equation $(x^2 - 1) \cdot (x^2 - 4) = 19$ have?

(A) 0   (B) 1   (C) 2   (D) 3   (E) 4

16. Which of the following circles allows the point $(3, 2)$ to be on it?

(A) $x^2 + (y - 1)^2 = 9$ (B) $(x - 1)^2 + (y - 1)^2 = 9$ (C) $x^2 + y^2 = 16$

(D) $x^2 + (y + 1)^2 = 36$ (E) $(x + 1)^2 + (y + 1)^2 = 25$

17. What is the area of the rhombus given in figure 3?

(A) 211.34

(B) 211.43

(C) 212.43

(D) 220.24

(E) 221.34

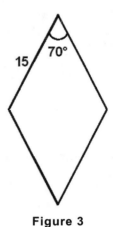

Figure 3
Figure not drawn to scale

GO ON TO THE NEXT PAGE ▶▶▶

Model Test 10

18. In figure 4, AB is the diameter and O is the center of the circle. If C is an arbitrarily selected point on the semicircular arc AB and $\cos\theta = \dfrac{5}{13}$, then $\tan\alpha = ?$

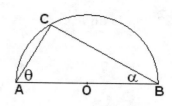

Figure 4
Figure not drawn to scale

(A) $\dfrac{5}{12}$ 　　(B) $\dfrac{5}{13}$ 　　(C) $\dfrac{12}{5}$ 　　(D) $\dfrac{13}{5}$ 　　(E) $\dfrac{13}{12}$

19. If $(1 - \sin x)(1 + \sin x) = 0.4$ then $\cos^2 x = ?$
(A) 0.3 　　　　(B) 0.4 　　　　(C) 0.5 　　　　(D) 0.6 　　　　(E) 0.8

20. Given in figure 5 is a circle having its center at O and a radius of length R. If the measure of θ is 40 degrees and R = 2.5 inches then what is the area of the minor sector AOB?

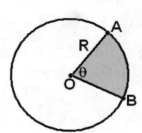

Figure 5

(A) 2.15 　　(B) 2.16 　　(C) 2.17 　　(D) 2.18 　　(E) 2.19

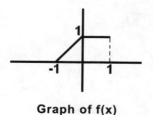

Graph of f(x)　　　**Graph of g(x)**

Figure 6

21. Based on the information given in figure 6 above, what is the relation between f(x) and g(x)?
(A) g(x) = f(-x-1) 　　(B) g(x) = f(-x+1) 　　(C) g(x) = f(-x)+1 　　(D) f(x) = g(-x-1) 　　(E) f(x) = g(-x)+1

GO ON TO THE NEXT PAGE ▶▶▶

Model Test 10

22. If α is an acute angle for which $\sin\alpha = \dfrac{3}{y}$ then $\tan\alpha$=?

(A) $\dfrac{\sqrt{y^2-9}}{3}$ (B) $\dfrac{3}{\sqrt{y^2-9}}$ (C) $\dfrac{\sqrt{y^2-9}}{y}$ (D) $\dfrac{y^2-9}{9}$ (E) $\dfrac{y}{3}$

23. The points (1, 2) and (5, -10) are symmetric with respect to which of the following points?

(A) (-3, 4)   (B) (-3, 14)   (C) (-2, 6)   (D) (3, -4)   (E) (9, -22)

24. What is the range of the relation given in figure 7?

(A) {-2, 2, 3}   (B) -6 < x < 5

(C) -6 ≤ x ≤ 5   (D) -6 ≤ x < 5

(E) {-6, -5,...,5}

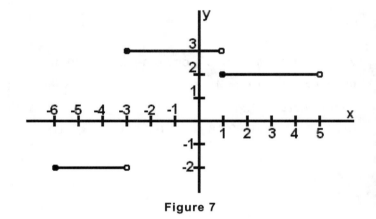

Figure 7

25. What is the range of the function given by $f(x) = 1 + \sqrt{1 + 5x^2}$?

(A) All real numbers

(B) All real numbers greater than 2

(C) All real numbers greater than or equal to 2

(D) All real numbers greater than or equal to $\dfrac{-1}{5}$

(E) All real numbers greater than or equal to $\dfrac{-1}{\sqrt{5}}$

GO ON TO THE NEXT PAGE ▶▶▶

Model Test 10

26. What is the slope of the line defined by x = 2t – 3 and y = -3t + 1?

(A) -3/2   (B) 3/2   (C) 2/3   (D) -2/3   (E) -6

27. Based on the data given in figure 8 what is the radius of the circle that passes through the vertices of the shaded rectangle?

(A) 5.5   (B) 6.0   (C) 6.5   (D) 13.0

(E) It cannot be determined from the information given.

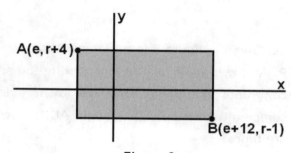

Figure 8
Figure not drawn to scale

28. How many of the following could be formed by the intersection of a cone and a plane?

   I. A point   II. A parabola   III. A circle   IV. A hyperbola

   V. An ellipse   VI. A triangle   VII. A line segment

(A) less than 4   (B) 4   (C) 5   (D) 6   (E) 7

29. In how many ways can six cards a, b, c, d, e and f be arranged so that card a is not in either end?

(A) 360   (B) 450   (C) 480   (D) 600   (E) 720

30. Which of the following functions has an inverse that does not equal itself?

(A) $f(x) = x$ (B) $f(x) = \dfrac{1}{x}$ (C) $f(x) = \dfrac{-3}{x}$ (D) $f(x) = -x + 3$ (E) $f(x) = \dfrac{x-1}{2}$

GO ON TO THE NEXT PAGE ▶▶▶

Model Test 10

31. What is the distance between one of the vertices of a cube and the center of the cube if the surface area of the cube is 24?

(A) $\sqrt{2}$ (B) $\sqrt{3}$ (C) $2\sqrt{2}$ (D) $2\sqrt{3}$ (E) 4

32. Given in figure 9 is a triangle ABC inscribed in the circle centered at P. A new triangle will be inscribed in the same circle such that it will be congruent to triangle ABC, one of its sides will be AB; but it will not completely overlap with triangle ABC. How many of such new triangles are there?

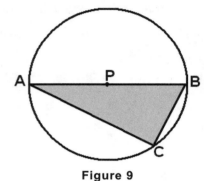

Figure 9

(A) 1   (B) 2   (C) 3   (D) 4

(E) more than 4

33. In a bag there are 5 identical white balls and 5 other balls each of which is unique. How many different sets of 5 balls can be selected from this bag?

(A) 2   (B) 5   (C) 6   (D) 31   (E) 32

34. The volume of a cube increases by 316 cubic inches when each side of the cube is increased by 4 inches. What is the length of a side of the original cube?

(A) 3   (B) 4   (C) 5   (D) 6   (E) 7

35. For how many pairs of positive integers (a, b) does the equation $a^2 - b^2 = 75$ hold?

(A) 0   (B) 1   (C) 2   (D) 3   (E) More than 3

GO ON TO THE NEXT PAGE ▶▶▶

Model Test 10

36. If x_1 and x_2 are the roots of a quadratic equation with rational coefficients then which of the following is not possible?

(A) $x_1 = x_2 = 1$

(B) $x_1 = x_2 = 3 - \sqrt{3}$

(C) $x_1 = 1 - \sqrt{2}$; $x_2 = 1 + \sqrt{2}$

(D) $x_1 = x_2 = \dfrac{2}{3}$

(E) $x_1 = 1 + i$; $x_2 = 1 - i$

37. Based on the information given in figure 10, x = ?

(A) 5.29   (B) 8.77   (C) 9.96   (D) 28   (E) 77

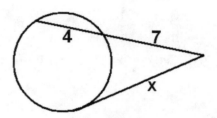

Figure 10
Figure not drawn to scale

38. How many four digit even numbers with distinct digits can be made by using the elements of the set {1, 2, 3, 4, 5, 6, 7, 8, 9}?

(A) 1344   (B) 2016   (C) 2919   (D) 3024   (E) 6561

39. 10% of the professors in **RUSH** academy participated in a chess competition where every participant had a match with every other participant exactly once and there were a total of 28 matches. How many professors are there in the academy?

(A) 60   (B) 70   (C) 80   (D) 90   (E) 100

40. If $x < 0 < y$ then which of the following is less than $\dfrac{x}{y}$ for all possible values of x and y?

(A) $-\dfrac{x}{3y}$ (B) $\dfrac{x}{3y}$ (C) $-\dfrac{3x^2}{y}$ (D) $\dfrac{3x}{y}$ (E) $\dfrac{x^2}{y}$

GO ON TO THE NEXT PAGE ▶▶▶

Model Test 10

41. If y=(x − 2)(x + 1), which of the following gives all values of x for which y > 0?

(A) − 1 < x < 2   (B) − 2 < x < 1   (C) x ≠ 2 and x ≠ -1   (D) x < − 2 or x > 1   (E) x < − 1 or x > 2

42. If point P has coordinates (x, y), where x = cos 45˚ and y = sin 135˚, then P lies on the graph of which of the following relations?

   I. |x| = |y|
   II. 2xy = 1
   III. $x^2 + y^2 = 1$

(A) I only   (B) II only   (C) I and III only   (D) II and III only   (E) I, II and III

43. What does the relation between x and y describe if the point (x, y) is equidistant from the point (3, -1) and the line x = -7?

(A) A point   (B) A line   (C) A circle   (D) A parabola   (E) An ellipse

44. The first and last terms of an arithmetic sequence with six terms are 1 and 10 respectively. Which of the following is one of the terms of this sequence?

(A) 2.9   (B) 4.8   (C) 6.4   (D) 6.6   (E) 8.4

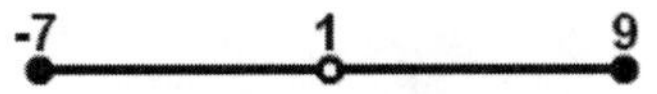

Figure 11

45. The graph given in figure 12 above is the solution set of

(A) |x-1| < 8   (B) |x-1| > 8   (C) |x-1| ≤ 8   (D) 0 < |x-1| < 8   (E) 0 < |x-1| ≤ 8

GO ON TO THE NEXT PAGE ► ► ►

46. If the semicircle given in figure 11 is rotated 90° about the x axis what would be the volume of the resulting three dimensional object?

(A) 359   (B) 457   (C) 718   (D) 1077   (E) 1437

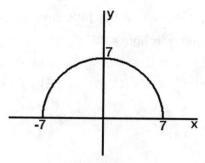

Figure 12
Figure not drawn to scale

47. Which of the following geometric figures has more than four lines of symmetry?

(A)	(B)	(C)	(D)	(E)
rectangle	trapezoid	circle	kite	square

48. In a production plant the cost C of production is defined by a linear function of the number of items produced. If the cost of producing 800 items is 2,000 dollars and that of 1400 items is 3,200 dollars then what will be the cost of producing 500 items?

(A) 1300   (B) 1400   (C) 1500   (D) 1600   (E) 1700

49. If $\log_3(x + 5) - \log_3(x - 5) = 2$ then x = ?

(A) 4.25   (B) 5.25   (C) 6.25   (D) 8.25   (E) 9.25

GO ON TO THE NEXT PAGE ▶▶▶

Model Test 10

50. Which of the following inequalities represents the points (x, y) given in figure 13?

(A) $y > |x - 4|$ (B) $y \geq |x - 4|$ (C) $y \geq |x + 4|$

(D) $y < |x - 4|$ (E) $y \leq |x - 4|$

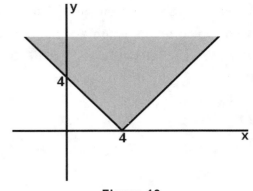

Figure 13

Figure not drawn to scale

S T O P

END OF TEST

(Answers on page 102 – Solutions on page 224)

Model Test 11

Test Duration: 60 Minutes

Directions: For each of the following problems, decide which is the **best** of the choices given. If the exact numerical value is not one of the choices, select the choice that best approximates this value. Then fill in the corresponding oval on the answer sheet.

Notes:

- A calculator will be necessary for answering some (but not all) of the questions in this test. For each question you will have to decide whether or not you should use a calculator. The calculator you use must be at least a scientific calculator; programmable calculators and calculators that can display graphs are permitted.

- The only angle measure used on this test is degree measure. Make sure your calculator is in the degree mode.

- Figures that accompany problems in this test are intended to provide information useful in solving the problems. They are drawn as accurately as possible **except** when it is stated in a specific problem that its figure is not drawn to scale.

- All figures lie in a plane unless otherwise indicated.

- Unless otherwise specified, the domain of any function f is assumed to be the set of all real numbers **x** for which **f(x)** is a real number.

Reference Information: The following information is for your reference in answering some of the questions in this test.

- Volume of a right circular cone with radius **r** and height **h**: $V = \frac{1}{3}\pi r^2 h$

- Lateral area of a right circular cone with circumference of the base **c** and slant height **l**: $S = \frac{1}{2}cl$

- Volume of a sphere with radius **r**: $V = \frac{4}{3}\pi r^3$

- Surface area of sphere with radius **r**: $S = 4\pi r^2$

- Volume of a pyramid with base area **B** and height **h**: $V = \frac{1}{3}Bh$

1. If a and b are distinct prime numbers, which of the following numbers must be odd?

(A) ab   (B) 4a+b-1   (C) a+b+5   (D) ab-1   (E) 2a-2b+1

2. If $3x^5 = 4$, then $5(3x^5)^2 = ?$

(A) 40   (B) 60   (C) 80   (D) 100

(E) None of the above

GO ON TO THE NEXT PAGE ▶▶▶

Model Test 11

3. The slopes of two lines that intersect at a single point can be

   I. equal

   II. reciprocals of each other

   III. negative reciprocals of each other

(A) I only   (B) II only   (C) III only   (D) II and III only   (E) I, II and III only

4. If $x^2 + y^2 = 5$ and $x^2 - y^2 = 3$, then $x^2 =$

(A) 1   (B) 2   (C) 3   (D) 4   (E) 5

5. If $\dfrac{5}{6}x = 0$, then $\dfrac{6}{5} - x = ?$

(A) 0   (B) $\dfrac{5}{6}$ (C) $\dfrac{6}{5}$ (D) $\dfrac{25}{36}$ (E) 1

6. What does the expression given by $x \cdot \left(\dfrac{1}{y} - \dfrac{1}{z} \right)$ equal?

(A) $\dfrac{x}{y-z}$ (B) $\dfrac{xz - xy}{yz}$ (C) $\dfrac{x}{yz}$ (D) $\dfrac{x}{z-y}$ (E) $\dfrac{1}{xy - xz}$

7. It is given that $0 < x + y < 17$ and x and y are distinct positive integers. If the greatest value of $x + y$ that is a prime number is n and the least negative value of $x - y$ that corresponds is m then $(m, n) = ?$

(A) (-1, 13)   (B) (-11, 13)   (C) (-13, 11)   (D) (13, -1)   (E) (13, -11)

GO ON TO THE NEXT PAGE ▶ ▶ ▶

Model Test 11

8. The median of ten consecutive odd integers is - 4. What is the sum of these integers?

(A) 5　　　　　　(B) 13　　　　　　(C) – 8　　　　　　(D) – 13　　　　　　(E) – 40

9. If all variables in the equations xy=35 and yz=45, represent integers greater than 1, then x+y+z=?

(A) 81　　　　　　(B) 45　　　　　　(C) 35　　　　　　(D) 21　　　　　　(E) 9

10. If $x^{2k-1} < 0$, where x is a real number and k is an integer, then

(A) x < 0　　　(B) x ≤ 0　　　(C) x > 0　　　(D) x ≥ 0　　　(E) x can be any real number

11. If it is given that $A@B = A^B - B^A$ and 2@C = 1 then C can be

(A) -2　　　　　　(B) -1　　　　　　(C) 3　　　　　　(D) 1　　　　　　(E) 2

12. A convex polygon has 9 diagonals. It is a

(A) pentagon (5 sides)　　　　　(B) hexagon (6 sides)　　　　　(C) heptagon (7 sides)

(D) octagon (8 sides)　　　　　(E) nonagon (9 sides)

13. Given that $z = t^3 - 1$ and $y = 3z^2 + 4$. If t = – 1 then y – z =?

(A) 18　　　　　　(B) -18　　　　　　(C) -2　　　　　　(D) -9　　　　　　(E) 9

GO ON TO THE NEXT PAGE ▶▶▶

Model Test 11

14. Given in figure 1, if ΔABC is reflected across line m, what will be the coordinates of the reflection of point B?

(A) (7, 8)   (B) (7, 3)   (C) (3, 7)   (D) (8, 7)

(E) None of the above

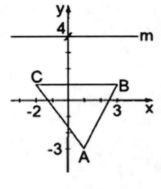

Figure 1

15. If in figure 2, BC is given to be x, then which of the following is equal to AC?

(A) $x \cdot \cos(\theta)$ (B) $x \cdot \sin(\theta)$ (C) $\dfrac{x}{\cos(\theta)}$ (D) $\dfrac{x}{\sin(\theta)}$ (E) $\dfrac{x}{\tan(\theta)}$

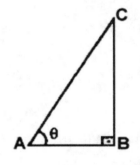

Figure 2

16. $f(x)= \sqrt{x-10}$; $g(x)= x^3 + x + 1$; $f(g(2))=$?

(A) ±1   (B) 11   (C) -1   (D) 1

(E) None of the above

17. In six years, Irem will be 4/5 as old as Sinan will be. In 15 years Irem will be 7/8 as old as Sinan will be. How old are they now?

(A) Irem is 2 and Sinan is 4.   (B) Irem is 8 and Sinan is 6.

(C) Irem is 6 and Sinan is 9.   (D) Irem is 7 and Sinan is 9.

(E) It cannot be determined from the given information.

GO ON TO THE NEXT PAGE ▶ ▶ ▶

Model Test 11

18. A quadrilateral is definitely a parallelogram if

   I. Two of its sides are equal in length and parallel.

   II. Its diagonals bisect each other.

   III. It has two pairs of supplementary angles.

(A) I only   (B) II only   (C) III only   (D) I or II   (E) I, II or III

19. If no Zulpies are Mulpies and all Mulpies are Tulpies, which of the following must be correct?

(A) All Tulpies are Mulpies.   (B) At least one Tulpy is not a Zulpy.

(C) Some Tulpies are Zulpies.   (D) A Tulpy that is not a Mulpy cannot be a Zulpy.

(E) Some Mulpies are Zulpies.

20. Melike and Dila went shopping. They spent half of what they had, plus $4 at the first store. At the second store, they spent half of what was left, plus $10. At the third store, they spent half of what was left. They spent the remaining $10 on hot candies. How much did they start with?

(A) $200   (B) $176   (C) $128   (D) $90

(E) It cannot be determined from the given information.

21. Which of the following points is closest to the point (1,-1)?

(A) (1, 7)   (B) (8, -1)   (C) (6, -2)   (D) (2, 7)   (E) (3, 5)

22. In a rectangle with sides of length 5 and 12, the angle that a diagonal makes with the shorter side is θ. Which of the following is incorrect?

(A) $1+\tan^2\theta=1/\cos^2\theta$ (B) $\cos\theta=5/13$ (C) $\tan\theta=5/12$

(D) $\sin^2(2\theta)+\cos^2(2\theta)=1$ (E) $\sin\theta=12/13$

GO ON TO THE NEXT PAGE ▶▶▶

Model Test 11

23. If $f(x) = x^2 - 9$ and $f(f(A))=0$ then A cannot be

(A) -3.5   (B) -2.4   (C) 0   (D) 2.4   (E) 3.5

Statement: The square of a number x is less than itself.

24. The set of all x values that satisfy the above statement are included in the set that contains

(A) the positive real numbers less than 1.   (B) the real numbers greater than 1.

(C) the negative real numbers greater than -1.   (D) the real numbers less than -1.

(E) none of the above.

25. Figure 3 shows a rectangle with sides of length 8 inches and 15 inches inscribed in a circle. What is the area of the shaded region in square inches?

(A) 100   (B) 103   (C) 106   (D) 107   (E) 109

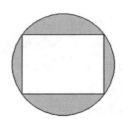

Figure 3
Figure not drawn to scale

26. The ratio of two numbers are 3 to 8 and their sum is -44. The lesser of the numbers

(A) is an odd multiple of 11   (B) is less than -22   (C) is an odd number

(D) is an even multiple of 11   (E) is greater than -22

GO ON TO THE NEXT PAGE ▶▶▶

27. Which of the following has more than one line of symmetry?

(A)   (B)   (C)   (D)   (E)

C E I T W

28. Which of the following intervals is a subset of all other intervals?

(A) $1 < x < 7$ (B) $2 \le x < 6$ (C) $3 \le x \le 4$ (D) $1 \le x < 7$ (E) $1 \le x \le 5$

29. In figure 4 given below, BE and DC are the bisectors of the angles $\angle ABC$ and $\angle ECF$ respectively. What is the relation between the measures of angles x and y?

(A) $y = 45° - x$ (B) $y = x/2$

(C) $y = 90° - 2x$ (D) $y = 2x$

(E) None of the above

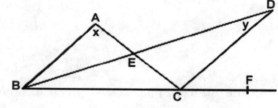

Figure 4

Figure not drawn to scale

30. What is the largest possible domain for the function $f(x) = \sqrt{x+1} + \dfrac{1}{x-1}$?

(A) $x > 1$ or $x \ne 1$ (B) $x \ge -1$ or $x \ne 1$ (C) $x > -1$ and $x \ne 1$

(D) $x > -1$ or $x \ne 1$ (E) $x \ge -1$ and $x \ne 1$

GO ON TO THE NEXT PAGE ▶ ▶ ▶

Model Test 11

31. The area of a triangle can be calculated by the formula $S = \sqrt{u(u-a)(u-b)(u-c)}$ where a, b, and c are the lengths of sides and u is the semi-perimeter of the triangle. Based on this information, what is the area of a triangle with sides of length 13, 14, and 15?

(A) 86   (B) 168   (C) 42   (D) 84   (E) 166

32. A regular octagon is inscribed in a square in such a way that four sides of the octagon coincide with the sides of the square as shown in figure 5. If the square has a side of length 100 inches, what is the length of one side of the octagon?

(A) $\dfrac{100}{2+\sqrt{2}}$ (B) $100(\sqrt{2}+2)$ (C) $\dfrac{100}{2-\sqrt{2}}$ (D) $100(-\sqrt{2}+2)$

(E) None of the above

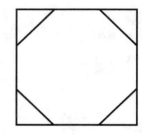

Figure 5

33. The right triangle ΔABC given in figure 6 is first reflected in the x axis and then rotated clockwise for 90° about the origin. The resulting figure is ΔA'B'C' where A', B' and C' are the images of points A, B, and C respectively. ΔA'B'C' can also be obtained by

(A) a single reflection of ΔABC in the line y = x.

(B) a single reflection of ΔABC in the line y = - x.

(C) a single reflection of ΔABC in the origin .

(D) a single clockwise rotation of 180° about the origin.

(E) None of the above.

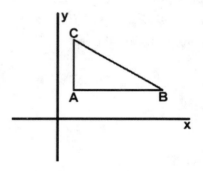

Figure 6

GO ON TO THE NEXT PAGE ▶▶▶

34. The curved path that encloses a square and two semicircles as shown in figure 7 is used to model a hockey field. If the length of the path is 100 inches, what is the area of the region that it encloses in square inches?

(A) 9.72   (B) 9,725   (C) 675.4   (D) 675,367

(E) None of the above

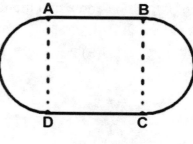

Figure 7

35. $g(x)$ is a function defined in terms of another function $f(x)$. If $g(-x) = -g(x)$ then $g(x)$ can be defined for all $f(x)$ as

    I. $g(x) = f^3(x)$

    II. $g(x) = f(-x) - f(x)$

    III. $g(x) = |f(-x)|$

(A) I only   (B) II only   (C) I and II only   (D) II and III only   (E) I, II and III

36. A palindrome is a number that reads the same forward as it does backward. How many 5 digit palindromes are there?

(A) 90   (B) 900   (C) 9000   (D) 90000   (E) 89999

x	y	1	2	3

old ID code

x	y	z	1	2

new ID code

37. In **RUSH** Gourmet, each dish is assigned an ID code that used to be formed with 2 letters followed by 3 digits. The new administration at **RUSH** changed the form of each ID code to 3 letters followed by 2 digits as indicated in the table above. Approximately how many more ID codes can be created now?

(A) 108,000   (B) 1,000,000   (C) 1,100,000   (D) 1,200,000   (E) 10,800,000

GO ON TO THE NEXT PAGE ▶ ▶ ▶

Model Test 11

38. If θ is one angle of a triangle then how many of the following values cannot be equal to cosθ?

I. -1   II. -0.25   III. 0   IV. 0.25   V. 1

(A) 1   (B) 2   (C) 3   (D) 4   (E) 5

39. For the trapezoid given in figure 8, AD=4, DC=6 and ∠ CBA measures 45°. If the trapezoid is rotated about AB, then what will be the volume of the resulting solid?

(A) 368   (B) 502   (C) 369   (D) 503   (E) 184

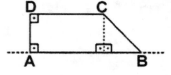

Figure 8

Figure not drawn to scale

40. The first three terms of an arithmetic sequence are 4t, 10t-1, and 12t+2. What is the numerical value of the fiftieth term?

(A) 74   (B) 79   (C) 244   (D) 249   (E) 254

41. What is the equation of the circle with the radius of 10 and center at (3,-1)?

(A) $(x-3)^2+(y+1)^2=10$ (B) $(x+3)^2+(y-1)^2=10$ (C) $(x-3)^2+(y+1)^2=100$

(D) $(x+3)^2+(y-1)^2=100$ (E) $(x+1)^2+(y-3)^2=100$

42. For a polynomial function P(x) what is the remainder when P(x+2) is divided by x+1?

(A) P(-3)   (B) P(-1)   (C) P(3)   (D) P(1)

(E) None of the above.

GO ON TO THE NEXT PAGE ▶▶▶

Model Test 11

43. What is the equation of the perpendicular bisector of the segment joining the points (1,2) and (3,6)?

(A) y-4=0.5(x-2)   (B) y-4= -2(x-2)   (C) y-4=2(x-2)   (D) y-4= -0.5(x-2)   (E) y-4= -5(x-2)

44. The two concentric circles given in figure 9 have radii of 5 and 3 and their centre is at O. If measure of angle ∠AOD is 72° then what is the perimeter of the shaded region?

(A) 10   (B) 12   (C) 14   (D) 16   (E) 20

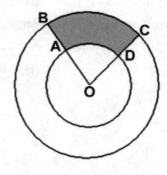

Figure 9

Figure not drawn to scale

45. If B(m, n) is a point on the line 3x-4y+1= 0, what is the minimum distance from the point A(3,5) to B?

(A) 2   (B) 4   (C) 6   (D) 8   (E) 10

46. Based on the data given in figure 10, what is the area between the rhombus and the square if the vertices A and C belong to both the square and the rhombus; one side of the square is 10 and angle ABC measures 40°?

(A) 40   (B) 41   (C) 42

(D) 43   (E) 44

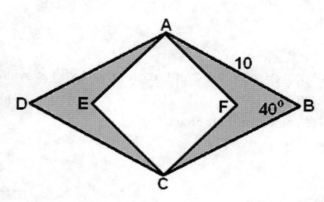

Figure 10

Figure not drawn to scale

GO ON TO THE NEXT PAGE ▶▶▶

Model Test 11

47. What is the acute angle between the lines $y = x + 3$ and $y = \sqrt{3}\,x + 2$?

(A) 15°   (B) 30°   (C) 45°   (D) 60°   (E) 75°

48. In a class, there are 23 boys and 19 girls. 10 of the boys and 7 of the girls have participated in the talent show. If a student is selected at random what is the probability that the student has participated in the talent show or is a boy?

(A) 5/7   (B) 10/21   (C) 10/17   (D) 5/21   (E) 10/23

49. As indicated in figure 10, a bigger square is divided into 25 unit squares that are identical in shape and a pattern is made by shading exactly one unit square in every row and column. How many such patterns can be made?

(A) 5   (B) 15   (C) 25   (D) 60   (E) 120

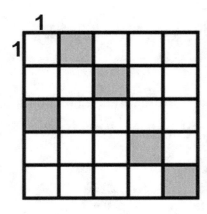

Figure 11

50. $f(x)=2x^2+12x+3$. If the graph of $f(x-k)$ is symmetric about the y axis, what is k?

(A) -3   (B) 3   (C) -15   (D) 15

(E) None of the above

S T O P

END OF TEST

(Answers on page 202 – Solutions on page 226)

Model Test 12

Test Duration: 60 Minutes

Directions: For each of the following problems, decide which is the **best** of the choices given. If the exact numerical value is not one of the choices, select the choice that best approximates this value. Then fill in the corresponding oval on the answer sheet.

Notes:

- A calculator will be necessary for answering some (but not all) of the questions in this test. For each question you will have to decide whether or not you should use a calculator. The calculator you use must be at least a scientific calculator; programmable calculators and calculators that can display graphs are permitted.

- The only angle measure used on this test is degree measure. Make sure your calculator is in the degree mode.

- Figures that accompany problems in this test are intended to provide information useful in solving the problems. They are drawn as accurately as possible **except** when it is stated in a specific problem that its figure is not drawn to scale.

- All figures lie in a plane unless otherwise indicated.

- Unless otherwise specified, the domain of any function f is assumed to be the set of all real numbers **x** for which **f(x)** is a real number.

Reference Information: The following information is for your reference in answering some of the questions in this test.

- Volume of a right circular cone with radius **r** and height **h**: $V = \frac{1}{3}\pi r^2 h$

- Lateral area of a right circular cone with circumference of the base **c** and slant height **l**: $S = \frac{1}{2}cl$

- Volume of a sphere with radius **r**: $V = \frac{4}{3}\pi r^3$

- Surface area of sphere with radius **r**: $S = 4\pi r^2$

- Volume of a pyramid with base area **B** and height **h**: $V = \frac{1}{3}Bh$

1. When b = 4, the point (1, 2) is on the line given by ax + b = 3y. a = ?

(A) 0   (B) 1   (C) 2   (D) -1   (E) -2

2. If p and r are two integers greater than 2, one of them being odd and the other being even, which of the following numbers must be even?

(A) $p^2 + r^2$ (B) $p^2 - r^2$ (C) $p^5 r^7$ (D) 3p - 35   (E) (p - r)!

GO ON TO THE NEXT PAGE ▶▶▶

Model Test 12

3. The rental cost of a motorboat is $10 per hour for each of the first 3 hours and $8 per hour for thereafter. Which of the following is an expression for the cost, in dollars, of renting this motorboat for n hours, in n > 3?

(A) 10 + 8n – 3   (B) 10 + 8(n-3)   (C) 10·3 + 8(n-3)   (D) 30 + (8n-3)   (E) 30 + 8n

4. |-2.8| - |5.4| + |-0.6|=

(A) -8.8   (B) -3.2   (C) -2   (D) 1.8   (E) 7.4

5. Rectangle ABCD in figure 1 consists of 9 congruent rectangles and it has a perimeter of 76 inches. What is its area in square inches?

(A) 38   (B) 40   (C) 180   (D) 360

(E) It cannot be determined from the given information

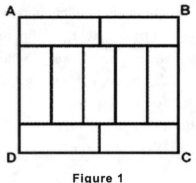

Figure 1

Figure not drawn to scale

6. The speed of a photocopy machine is n minutes per page in black & white mode and m minutes per page in color mode. How many hours, in terms of n and m would it take to copy a document that consists of 50 black & white pages and 60 pages in color?

(A) $\dfrac{5n}{6}+m$ (B) $\dfrac{5m}{6}+n$ (C) $\dfrac{5n+6m}{n+m}$ (D) 50n + 60m   (E) 110(n+m)

7. If a=p·b and c=r·d where all variables represent positive integers greater than 1, which one(s) of the following are always true?

I. $\dfrac{a}{b}>1$ II. a > c   III. $\dfrac{d}{c}>1$

(A) None   (B) I only   (C) II only   (D) I and II only   (E) II and III only

GO ON TO THE NEXT PAGE ▶▶▶

Model Test 12

8. Right triangle ABC is inscribed in the semicircle whose radius is 4 as given in the figure 2. If $\angle ABC$ measures 30° then what is the area of the shaded region?

(A) 11.2   (B) 11.3   (C) 22.5   (D) 22.6   (E) 36.4

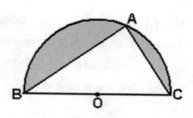

Figure 2
Figure not drawn to scale

9. What is the relation between angles a, b, and c given in figure 3?

(A) a + b = 180° − c

(B) a = b + c

(C) c = a + b

(D) a + c = 180° − b

(E) b = a + c

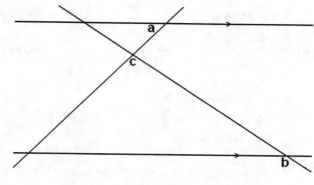

Figure 3

10. The regular pentagon ABCDE and the square GFAB have a common edge AB as is given in figure 4. If x is the measure of angle BCG then x = ?

(A) 90°   (B) 81°   (C) 78°   (D) 72°   (E) 54°

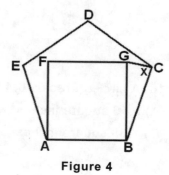

Figure 4

11. $\log_3 84$ is between what pair of consecutive integers?

(A) 1 and 2   (B) 2 and 3   (C) 3 and 4   (D) 4 and 5   (E) 5 and 6

12. A line has parametric equations x=2-t and y=t+2, where t is the parameter. What is the y − intercept of this line?

(A) -2   (B) 0   (C) t   (D) 2   (E) 4

GO ON TO THE NEXT PAGE ▶▶▶

13. At a certain sports competition every entrant is scored by six referees each of whom rates the entrant with a score selected from the set {1, 2, 3, 4, 5, 6, 7}. The cumulative scores of Ali, Berk, Cem, Deniz, and Emre are 24, 29, 34, 39, and 41 respectively. How many of these five entrants must have received 7 from one or more referees?

(A) 1   (B) 2   (C) 3   (D) 4   (E) 5

14. If the lines given by $\begin{cases} -3x + y = 6 \\ 2x - ky = 3 \end{cases}$ are perpendicular, then k=

(A) -6   (B) -3   (C) -1/6   (D) 3   (E) 6

15. What is the solution set of the inequality given by $\dfrac{x-1}{x} \geq 1$?

(A) x < 0   (B) -1 < x < 0   (C) 0 < x < 1   (D) x > 0   (E) x > 1

16. A circle of radius 0.42 inches is inscribed in a square. What is the approximate area of this square?

(A) 0.18 square inches   (B) 0.71 square inches   (C) 0.56 square inches

(D) 1.68 square inches   (E) 3.36 square inches

17. Graph of f(x) is given in figure 5. If g(x) = -f(-x) then how can the graph of g(x) be obtained?

(A) by reflecting the graph of f(x) across origin.

(B) by reflecting the graph of f(x) across the x axis.

(C) by reflecting the graph of f(x) across the y axis.

(D) by reflecting the graph of f(x) across the line y = x.

(E) by reflecting the graph of f(x) across the line y = -x.

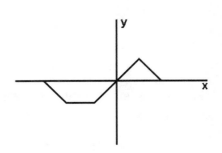

Figure 5

GO ON TO THE NEXT PAGE ▶ ▶ ▶

Model Test 12

18. Where defined $\left(\dfrac{x^2 + 2x - 3}{x - 2}\right)\left(\dfrac{x^2 - x - 2}{x + 3}\right)$ equals which of the following expressions?

(A) x -1   (B) x + 1   (C) 2x   (D) x^2-1   (E) $(x-1)^2$

19. In the figure 6, a regular hexagon with a side of length 4 is given. What are the coordinates of point A?

(A) $(4, -2\sqrt{3})$ (B) $(-4, -3\sqrt{3})$ (C) $(-2\sqrt{3}, -4)$ (D) $(-4, -2\sqrt{3})$ (E) $(-4, -2)$

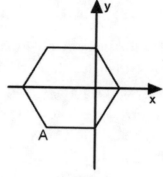

Figure 6

20. If two vertices of a square with a side length of $3\sqrt{2}$ in the xy-plane are at the origin and on the line y=x, which of the following cannot be another vertex of the square?

(A) (-6, 6)   (B) (-3, 3)   (C) (3, 3)   (D) (0, 6)   (E) $(3\sqrt{2}, 0)$

21. In an arithmetic sequence the 4th term is n and nth term is 4; what is the 2nd term?

(A) -2   (B) n-2   (C) 0   (D) 2   (E) n+2

22. Each of the 18 students in the senior class plays at least one of the sports of basketball, volleyball and football. If 5 students play all three of them and 8 students play exactly two, how many of these students play only one?

(A) 5   (B) 8   (C) 10   (D) 13   (E) 31

GO ON TO THE NEXT PAGE ▶▶▶

Model Test 12

23. If in figure 8 ABCD is a square and EAD is an equilateral triangle, what is the degree measure of angle AFC?

(A) 105   (B) 110   (C) 115

(D) 120   (E) 135

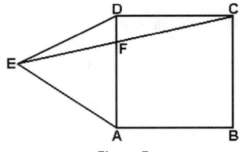

Figure 7

Figure not drawn to scale

24. Parallelogram ABCD in figure 7 consists of 8 congruent right triangles each of which is similar to triangle PRS with sides of length 3, 4, and 5. If the parallelogram has a perimeter of 64 inches then what is the area of the parallelogram?

(A) 128   (B) 144   (C) 192

(D) 256   (E) 384

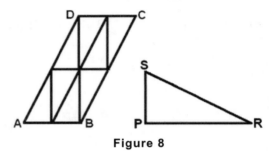

Figure 8

25. In right triangle ABC in the figure 9, tan $\angle$B is

(A) -6/5   (B) -5/6   (C) 1   (D) 5/6   (E) 6/5

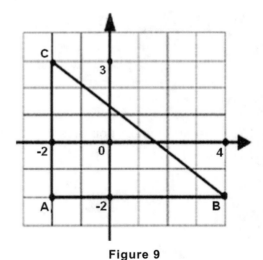

Figure 9

26. If f(x)= $\sqrt{4-x^2}$ which of the following numbers is not in the domain of f?

(A) -1   (B) 0   (C) 1   (D) 2   (E) 3

GO ON TO THE NEXT PAGE ▶ ▶ ▶

27. Which of the following diagrams is the best representation of the solution set for the system of inequalities given by y<−2x+1 and −1< x ≤ 2?

(A)
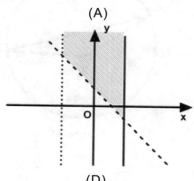

(B)
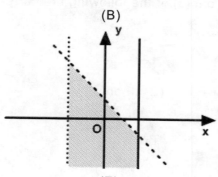

(C)
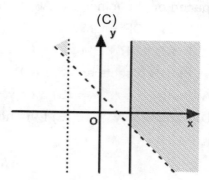

(D)
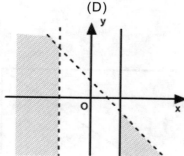

(E)
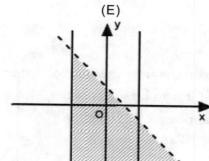

28. If f(x)= $\dfrac{1}{(x-1)^2}$ for all x≠1, which of the following statements must be true?

    I. f(0) = f(2)

    II. f(-1) = f(3)

    III. f(-2) = f(2)

(A) I only   (B) II only   (C) I and II only   (D) I and III only   (E) II and III only

29. If f(x)= x-2 and g(x)= $\dfrac{x^2-4}{x-2}$, how are the graphs of f and g related?

(A) They are exactly the same.   (B) They are the same except x = -2.

(C) They are the same except x = 2.   (D) They have no points in common.

(E) They have the same shape but only a finite number of points in common.

GO ON TO THE NEXT PAGE ►►►

Model Test 12

30. In figure 10, O is the center and points A, R, M, and B are equally spaced on the minor arc $\overset{\frown}{AB}$. Which one(s) of the following are correct?

  I. AB < OR + OM

  II. AB ≤ BR + BM

  III. OR bisects angle ∠AOM.

(A) I only   (B) II only   (C) III only

(D) I and II only   (E) I, II and III

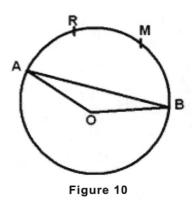

Figure 10

31. ABCD in figure 11 is a rectangular sheet of raw fabrics with dimensions of 20 and 16 feet and it is to be folded along line AE so that point B coincides CD at point F. CE must be

 (A) 12   (B) 6   (C) 10   (D) 16   (E) 8

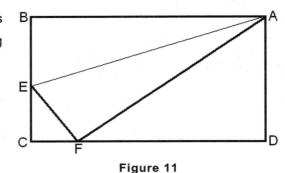

Figure 11
Figure not drawn to scale

32. A circle is inscribed in triangle ABC given in figure 12, touching segment BC in P, as shown in figure. If AB=10, BC=8 and CA=7, then PC is

 (A) 2   (B) $\dfrac{5}{2}$ (C) 3   (D) $\dfrac{7}{2}$ (E) 4

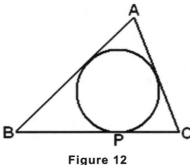

Figure 12
Figure not drawn to scale

33. If 12 yellow and 15 blue cubes that are identical in shape and dimensions are assembled to make one big cube. What is the greatest ratio of the yellow surface to the blue surface the resulting cube can have?

(A) 4/5   (B) 5/4   (C) 4/9   (D) 16/11   (E) 11/16

GO ON TO THE NEXT PAGE ▶▶▶

Model Test 12

34. Figure 13 shows the graph of a circle given by $(x+3)^2 + y^2 = 9$. If the circle is rotated 180° about the x-axis, what will be the volume of the solid that results?

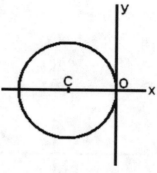

Figure 13

(A) 9π   (B) 12π   (C) 18π   (D) 24π   (E) 36π

35. The cube of a number n is decreased by 4. If the square root of that result is a two-digit palindromic number that reads the same forward as it does backward, then n can be

(A) 2   (B) 4   (C) 5   (D) 6   (E) 7

36. It is given that a and b are both real numbers and $a - 4 + bi = b + 2ai$ where $i^2 = -1$. What is the value of b?

(A) -4   (B) -8   (C) 2   (D) 4   (E) 8

37. If line d is the perpendicular bisector of the line segment with endpoints (-3,0) and (0,3), what is the slope of line d?

(A) -3   (B) -1   (C) 0   (D) 1   (E) 3

38. Defne and Yasmin left for a camp on the same day. Defne took 26 biscuits and ate 2 of them per day. Yasmin took 32 biscuits and didn't eat any biscuits for the first 3 days, but thereafter she ate 4 of them per day. In order to find the day when they have equal number of biscuits, which of the following equations can be used?

(A) $26 - 2n = 32 - 4n - 3$ (B) $32 - 2n = 26 - 4 \cdot (n - 3)$ (C) $26 - (n - 3) = 32 - 4 \cdot 2n$

(D) $32 - 2n = 26 - 4n - 3$ (E) $26 - 2n = 32 - 4 \cdot (n - 3)$

GO ON TO THE NEXT PAGE ▶▶▶

Model Test 12

39. Figure 14 is the graph of y=|f(x)|. Which of the following could be the graph of y= f(x)?

(A)

(B)

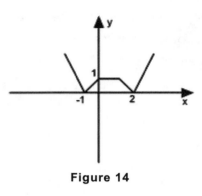

Figure 14

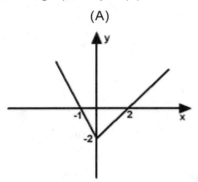

(C)

(D)

(E)

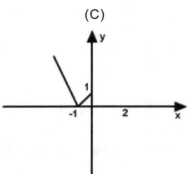

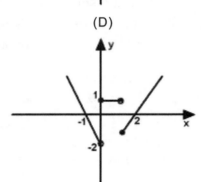

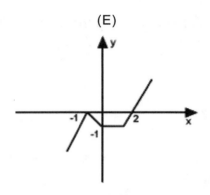

40. The circle in figure 15 has center at O. If AB is secant and AC is tangent to the circle then which of the following correctly gives the relation between the degree measures of α and θ where the measure of $\angle$DBC is 40 degrees?

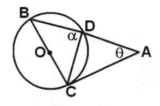

Figure 15
Figure not drawn to scale

(A) $\alpha < \theta - 40°$　　　(B) $\alpha > \theta + 40°$　　　(C) $\alpha = \theta + 40°$

(D) $\alpha < 2\theta - 50°$　　　(E) $\alpha + \theta = 180°$

41. If the measure of one angle of a rhombus is 60°, then what is the ratio of the length of its longer diagonal to the length of one of its sides?

(A) $\dfrac{1}{2}$　　　　(B) $\dfrac{\sqrt{3}}{2}$　　　　(C) $\sqrt{3}$　　　　(D) 2　　　　(E) $2\sqrt{3}$

GO ON TO THE NEXT PAGE ▶▶▶

Model Test 12

42. In triangle PRS, side PR is 10 and the angle S opposite this side is 60°. What is the radius of the circle that passes through the vertices of this triangle?

(A) 5   (B) 5.8   (C) 7.1   (D) 10   (E) It cannot be determined from the information given.

43. If $\log_3 2 = x$, then $\log_8 3 =$

(A) 3x   (B) x/3   (C) x^3 (D) $1/x^3$ (E) 1/(3x)

44. Given in figure 16, sides AB and AC of the triangle ABC have lengths of 6 and 7 respectively. If D is the midpoint of side BC then length of AD cannot be

(A) 6.5   (B) 5.3   (C) 4.2   (D) 3.7   (E) 0.9

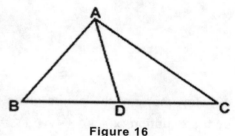

Figure 16
Figure not drawn to scale

45. If the total surface area of a rectangular solid A with dimensions of 3, 5, and n is less than the total surface area of a rectangular solid B with dimensions 2, 3, and 2n, what is the least possible integer value of n?

(A) 4   (B) 5   (C) 6   (D) 7   (E) 8

46. In the xy plane, what is the area of the triangle whose vertices are $(-2, \sqrt{2})$, $(3, 6)$ and $(5\sqrt{2}, \sqrt{2})$?

(A) 15.21   (B) 11.63   (C) 20.80   (D) 27.21   (E) 33.63

GO ON TO THE NEXT PAGE ▶▶▶

Model Test 12

47. If in figure 17 U is the set of quadrilaterals and A and B are the sets of parallelograms and rectangles respectively, B∩C and C can be respectively the sets of

(A) squares and kites
(B) kites and trapezoids
(C) rhombuses and kites
(D) squares and rhombuses
(E) rhombuses and squares

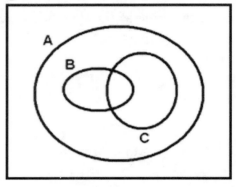

Figure 17

48. How many distinct positive divisors does $N = p \cdot q^4$ have if p and q represent distinct prime numbers greater than 7?

(A) 5   (B) 6   (C) 8   (D) 9   (E) 10

49. How many of the points given in the table are inside the circle given by the equation $(x - 1)^2 + (y + 1)^2 = 9$?

(A) None   (B) 1   (C) 2   (D) 3   (E) More than 3

Point Label	Coordinates
A	(1, 1)
B	(4, 0)
C	(0, 4)
D	(1, 2)
E	(0, 0)
F	(-2, 1)
G	(-2, 0)

50. Which of the following choices best describes the points (x, y) on the Cartesian plane that satisfy the inequalities given by: $x \geq -2$; $y \leq 10$; $y \geq 3x - 2$; and $y \leq 3x+1$?

(A) points on a triangle

(B) points on a trapezoid

(C) points inside a trapezoid

(D) points inside and on a triangle

(E) points inside and on a trapezoid

S T O P

END OF TEST

(Answers on page 202 – Solutions on page 228)

Model Test 13

Test Duration: 60 Minutes

Directions: For each of the following problems, decide which is the **best** of the choices given. If the exact numerical value is not one of the choices, select the choice that best approximates this value. Then fill in the corresponding oval on the answer sheet.

Notes:

- A calculator will be necessary for answering some (but not all) of the questions in this test. For each question you will have to decide whether or not you should use a calculator. The calculator you use must be at least a scientific calculator; programmable calculators and calculators that can display graphs are permitted.

- The only angle measure used on this test is degree measure. Make sure your calculator is in the degree mode.

- Figures that accompany problems in this test are intended to provide information useful in solving the problems. They are drawn as accurately as possible **except** when it is stated in a specific problem that its figure is not drawn to scale.

- All figures lie in a plane unless otherwise indicated.

- Unless otherwise specified, the domain of any function f is assumed to be the set of all real numbers **x** for which **f(x)** is a real number.

Reference Information: The following information is for your reference in answering some of the questions in this test.

- Volume of a right circular cone with radius **r** and height **h**: $V = \dfrac{1}{3}\pi r^2 h$

- Lateral area of a right circular cone with circumference of the base **c** and slant height **l**: $S = \dfrac{1}{2}cl$

- Volume of a sphere with radius **r**: $V = \dfrac{4}{3}\pi r^3$

- Surface area of sphere with radius **r**: $S = 4\pi r^2$

- Volume of a pyramid with base area **B** and height **h**: $V = \dfrac{1}{3}Bh$

1. Which of the following is different from the others?

(A) $m = -n^6$ (B) $-m = n^6$ (C) $-m = (-n)^6$ (D) $m = (-n)^6$ (E) $m + n^6 = 0$

2. x and y are real numbers such that $0 < y^x < y < 1$. Which of the following includes all possible values of x?

(A) $x < -1$ (B) $-1 < x < 0$ (C) $x < 0$ (D) $0 < x < 1$ (E) $x > 1$

GO ON TO THE NEXT PAGE ▶▶▶

3. For the equations $\sqrt{2}x + 2y = 1$ and $x - \sqrt{2}y = \sqrt{2}$ find x?

(A) $-\dfrac{1}{4}$ 　　　　(B) $\dfrac{\sqrt{2}}{2}$ 　　　　(C) $3\sqrt{2}$ 　　　　(D) $\dfrac{3\sqrt{2}}{4}$ 　　　　(E) $\dfrac{3\sqrt{2}}{2}$

4. In Emre's bookshelf there are m mathematics books and p physics books. When Emre buys n more mathematics books he will have twice as many mathematics books as physics books. Which of the following gives the relation between m, n and p correctly?

(A) m + n = 2p 　　　(B) m − n = 2p 　　　(C) m + p = 2n 　　　(D) p + n = 2m 　　　(E) p − n = 2m

5. The graph of the equation y = 2x + 3 lies in quadrants

(A) I and II only 　　　　　　(B) I and III only 　　　　　　(C) II and II only

(D) I, II and III only 　　　　　(E) II, III and IV only

6. It is given that pa + qb = 7 and pb + qa = 8. If p + q = 3 then $a^2 + b^2 + 2ab$ = ?

(A) 3 　　　　　(B) 4 　　　　　(C) 5 　　　　　(D) 16 　　　　　(E) 25

7. What is the area of rectangle ABCD given in figure 1?

(A) 10 　　(B) 12 　　(C) 15 　　(D) 18 　　(E) 21

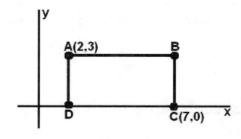

Figure 1
Figure not drawn to scale

GO ON TO THE NEXT PAGE ▶▶▶

Model Test 13

8. Which of the following is not a rational number?

(A) 4!   (B) 3.212121...   (C) $\sqrt{81}$ (D) $\frac{\pi}{2}$ (E) $\frac{\log 100}{5}$

9. If the solution set of the system of equations given is $\{(1, -1)\}$ then what is the value of ab?

$$(a+1)x - by - 3 = 0$$
$$(a-2)x + (b+3)y - 5 = 0$$

(A) -24   (B) -4   (C) 2   (D) 6   (E) 24

10. At Showtime Theaters the price of a student ticket is $2 and the price of an adult ticket is 1.5 as much as the price a student ticket. On a particular day $2,300 was collected from a sale of 900 tickets. How many of these were adult tickets?

(A) 300   (B) 400   (C) 500   (D) 600   (E) 700

11. If $\frac{2 \cdot \sin^2 x + 2 \cdot \cos^2 x}{\cos x} = 5$ when $0 < x < 90°$ then what is the measure of x rounded to the nearest degree?

(A) 23   (B) 24   (C) 32   (D) 66   (E) 68

12. For which of the following sets of numbers is the mean greater than the median?

(A) {101, 102, 103, 104, 105}   (B) {100, 102, 103, 104, 106}

(C) {101, 103, 103, 103, 105}   (D) {101, 102, 103, 104, 105}

(E) {101, 102, 103, 104, 106}

GO ON TO THE NEXT PAGE ▶▶▶

Model Test 13

13. The cost of a certain computer at **RUSH** store is \$c and the rate of profit is p%. At the end of season sale, the computer is discounted by d%. If there is also a sales tax of t% on the price of the computer, what is the total money that a customer will pay for buying the computer during the season?

(A) $c \cdot \left(1 + \dfrac{p}{100}\right) \cdot \left(1 - \dfrac{d}{100}\right) \cdot \left(1 + \dfrac{t}{100}\right)$

(B) $c \cdot \left(1 + \dfrac{p}{100}\right) \cdot \left(1 + \dfrac{t}{100}\right)$

(C) $c \cdot \dfrac{p}{100} \cdot \dfrac{t}{100}$

(D) $c \cdot \left(1 - \dfrac{p}{100}\right) \cdot \left(1 + \dfrac{d}{100}\right) \cdot \left(1 - \dfrac{t}{100}\right)$

(E) $c \cdot \dfrac{p}{100} \cdot \dfrac{d}{100} \cdot \dfrac{t}{100}$

14. Which of the following represents the statement, "When the square of the sum of twice x and three times y is subtracted from the cube of z, the result is p"?

(A) $z^3 - (2x + 3y)^2 = p$

(B) $z^3 - (4x^2 + 9y^2) = p$

(C) $(2x)^2 + (3y)^2 = p + z^3$

(D) $z^3 + (2x + 3y)^2 = p$

(E) $(2x + 3y)^2 - z^3 = p$

15. Which of the following is the smallest?

(A) 70%

(B) $\sqrt{0.7}$

(C) $(0.7)^{-1}$

(D) $(0.7)^2$

(E) $(0.7)^{\frac{1}{3}}$

16. If 3 + 4i is one root of a quadratic equation given by $x^2 - Px + Q = 0$ where P and Q are real numbers then P is

(A) 0

(B) 3

(C) 6

(D) – 3

(E) – 6

17. An arc that intercepts a central angle of 225° is what percent of the whole circle?

(A) 60%

(B) 62.5%

(C) 65%

(D) 67.5%

(E) 70%

GO ON TO THE NEXT PAGE ▶▶▶

18. If the right triangle given in figure 2 is rotated about the x axis for 360° what will be the volume of the resulting solid?

(A) 32.98   (B) 32.99   (C) 65.97   (D) 153.93   (E) 153.94

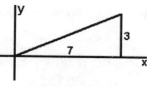

Figure 2

19. In figure 3, the upper and lower bases of the rectangle are divided into 4 and 3 equal segments respectively. The shaded area is what portion of the whole rectangle?

(A) 1/8   (B) 1/6   (C) 7/24   (D) 15/48
(E) It cannot be determined from the information given.

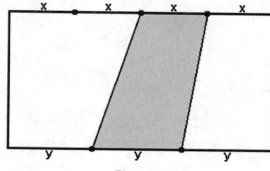

Figure 3

20. Which of the following can be the equation of the circle given in figure 4?

(A) $(x - 4)^2 + (y + 6)^2 = 4$ (B) $(x - 6)^2 + (y + 4)^2 = 16$
(C) $(x - 4)^2 + (y + 6)^2 = 16$ (D) $(x + 6)^2 + (y - 4)^2 = 16$
(E) $(x - 4)^2 + (y + 6)^2 = 36$

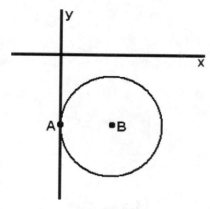

Figure 4

21. An object is projected vertically upwards so that t seconds later its height in meters is given by $h(t) = 100 + 20t - 5t^2$. The object reaches its maximum height when

(A) t = 2   (B) t = 3   (C) t = 4   (D) t = 5   (E) t = 6

22. If the distance between point P(m, n) and the origin is 5 inches then what is the distance between origin and the point whose coordinates are given by (4m, 4n)?

(A) 5   (B) 20   (C) 30   (D) 40   (E) 80

GO ON TO THE NEXT PAGE ▶▶▶

Model Test 13

1. P = 1 and Q = 1
2. If Q > 2000 then go to 5 otherwise go to 3
3. Replace Q by 3Q and P by P+1.
4. Go to 2
5. Print P

23. If the above instructions are carried out, what number will be printed?

(A) 7   (B) 8   (C) 9   (D) 10   (E) 11

24. The curve given by $y = x^3 - x^2$ is reflected across x-axis to get another curve given by $y = f(x)$; $f(x) = ?$

(A) $f(x) = x^3 + x^2 + 1$ (B) $f(x) = -x^3 + x^2$ (C) $f(x) = x^3 + x^2$

(D) $f(x) = (x+1)^3 + (x+1)^2$ (E) $f(x) = -(x^3 + x^2)$

25. Which of the following must be rotated at least 360° about a horizontal axis to give the three dimensional object given in figure 5?

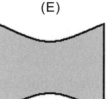

Figure 5

(A)

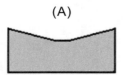

(B)

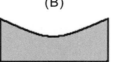

(C)

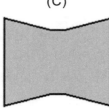

(D)

(E)

26. If $i = \sqrt{-1}$ then what is the distance between the numbers -2+2i and 6-4i in the complex plane?

(A) 6   (B) 8   (C) 10   (D) 12   (E) 14

GO ON TO THE NEXT PAGE ▶▶▶

27. In figure 6, the measure of θ is 2 degrees; how high is the building to the nearest yard?

(A) 78   (B) 79   (C) 87   (D) 92   (E) 97

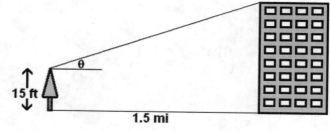

Figure 6

Figure not drawn to scale

28. A gambler will play a game that he will have to choose randomly from the list given in the table to the right. Assuming that he is equally likely to choose a game, what is the probability that he will choose game E and win?

(A) 0.07   (B) 0.13   (C) 0.35   (D) 0.65   (E) 0.7

Game	Probability of losing
A	0.45
B	0.15
C	0.25
D	0.30
E	0.35

29. What is the equation of the line whose x and y intercepts are P and -Q respectively?

(A) $\dfrac{x}{P}+\dfrac{y}{Q}=1$ (B) $\dfrac{x}{Q}+\dfrac{y}{P}=1$ (C) $\dfrac{x}{P}-\dfrac{y}{Q}=1$ (D) $\dfrac{x}{Q}-\dfrac{y}{P}=1$ (E) $\dfrac{x}{P}+\dfrac{y}{Q}=-1$

30. What is the area of the isosceles triangle given in figure 7?

(A) 45   (B) 81   (C) 86.4   (D) 86.5   (E) 172.9

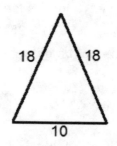

Figure 7

GO ON TO THE NEXT PAGE ▶▶▶

31. Based on the information given in figure 8, what is the length of x?

(A) 25   (B) 18   (C) 17   (D) 15   (E) 13

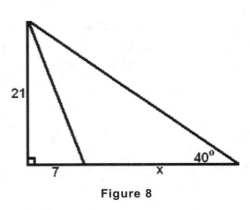

Figure 8

32. The pie chart given in figure 9 shows the classification of the talent show participants in **RUSH** Academy. The measure of the central angles x, y and z are 80°, 120°, 100° and of the students who attended the talent show, the number of seniors was 24 more than that of the sophomores. How many freshmen participated the talent show altogether?

(A) 36   (B) 48   (C) 60

(D) 72   (E) 216

Talent Show Participation in RUSH Academy

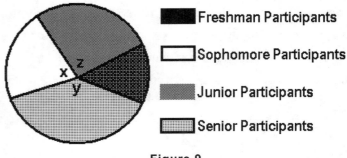

Figure 9

33. In figure 10, CB is tangent to the circle at point B and CD is tangent to the circle at point D. If A is the center of the circle and BCD is a right angle then what is the area of quadrilateral ABCD in square feet given that BD is 18 inches long?

(A) 162   (B) 2.25   (C) 1.125   (D) 1.15

(E) It cannot be determined from the information given.

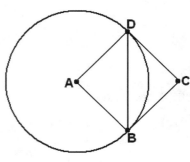

Figure 10

34. A unitary number is one such that when its digits are added, when the digits of the resulting number are added, and so on, the final result is 1. For example, 8767 is a unitary number because 8+7+6+7=28; 2+8=10; and 1+0=1. How many three digit unitary numbers are there?

(A) 98   (B) 99   (C) 100   (D) 101   (E) 102

GO ON TO THE NEXT PAGE ▶ ▶ ▶

Model Test 13

35. A line m passes through the points (-2, 1) and (3, 6) and another line n that passes through (2, 1) makes an angle of 135° with the positive x axis. What is the point of intersection of the lines m and n?

(A) (0, 3)　　　　(B) (-1, 2)　　　　(C) (1, 4)　　　　(D) (3, 0)　　　　(E) (2, -1)

36. If f(x) = 3x + 3 then f(x − 1) looks like

(A) 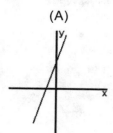   (B) 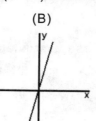   (C) 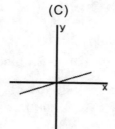   (D) 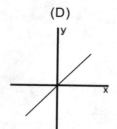   (E)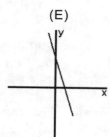

37. If $f(x) = x^2 - 1$ then $f^{-1}(x) = $?

(A) $f^{-1}(x) = \sqrt{x+1}$　　　　(B) $f^{-1}(x) = (x+1)^2$　　　　(C) $f^{-1}(x) = \pm\sqrt{x+1}$

(D) $f^{-1}(x) = -\sqrt{x+1}$　　　　(E) $f^{-1}(x) = -(x+1)^2$

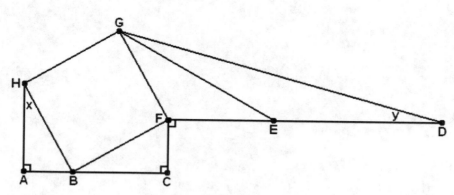

Figure 11

38. In figure 11 above, BFGH is a square; FGE and EGD are isosceles triangles whose vertices are at F and E respectively. If the measure of x is 30° then what is the measure of y in degrees?

(A) 15°　　　　(B) 20°　　　　(C) 25°　　　　(D) 30°　　　　(E) 35°

GO ON TO THE NEXT PAGE ▶▶▶

Model Test 13

39. If the equation of the curve given in figure 12 is |x + 5| + |y − 5| + k = 0 then k is

(A) − 5   (B) − 4   (C) 0   (D) 4   (E) 5

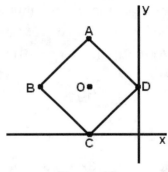

Figure 12

40. A certain transformation maps every point (x, y) to a point whose coordinates are given by (-2x, y). Which of the following gives all points (x, y) in the Cartesian plane that the transformation does not change?

(A) all points on the line y = x   (B) the x axis   (C) the origin

(D) all points on the line y = -x   (E) the y axis

41. How many shaded squares are there in the n'th pattern given in figure 13 above?

(A) $\dfrac{n^2+n+2}{2}$ (B) $\dfrac{n^2+2n+1}{2}$ (C) $\dfrac{2n^2+n+1}{2}$

(D) $\dfrac{n^2-n+2}{2}$ (E) $\dfrac{n^2+n-2}{2}$

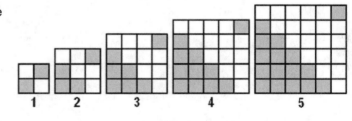

Figure 13

42. If $2^{x+1} = 0.25$, then $\log_3(4-2x) = ?$

(A) 2.095   (B) 2.096   (C) -3   (D) 5.092   (E) 5.093

GO ON TO THE NEXT PAGE ► ► ►

Model Test 13

43. In figure 14, both circles are centered at P and BCDE is a rectangle whose vertices lie on the greater circle If PA = 5 inches and PB = 13 inches then what is the area of the shaded region correct to the nearest ten square inches?

(A) 150　　　　(B) 160　　　　(C) 170　　　　(D) 180

(E) It cannot be determined from the information given.

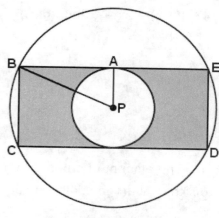

Figure 14

44. If two planes W and V are parallel, which of the following must be true?

I. If a line intersects W, it also intersects V.

II. If a line is parallel to W, it is also parallel to V.

III. If a line is perpendicular to W, it is also perpendicular to V.

(A) II only　　　(B) III only　　　(C) I and II only　　　(D) I and III only　　　(E) II and III only

45. If the graph of f(x) is given in figure 15, then $f(f(-2)+f(1)+1)=?$

(A) -2　　　　(B) -1　　　　(C) 0　　　　(D) 1　　　　(E) 2

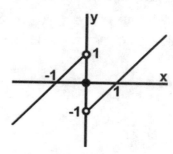

Figure 15

"If 2n + 1 is prime then 2n − 1 and 2n + 5 are also prime."

46. Which of the following values of n results in a counter example of the above statement?

(A) 3　　　　　(B) 6　　　　　(C) 8　　　　　(D) 9　　　　　(E) 21

GO ON TO THE NEXT PAGE ▶▶▶

Model Test 13

47. If f(x) = 2x − 1 and f(g(x)) = 3x + 4 then g(x)=?

(A) $\dfrac{3x+5}{2}$ (B) $\dfrac{5x+3}{3}$ (C) $\dfrac{3x+11}{2}$ (D) $\dfrac{3x-5}{3}$ (E) $\dfrac{2x-11}{3}$

48. The amount of crops C in a certain year that a field will yield in tons per square meters is a function of the average daily amount of rain R in cubic meters in that year and it is given by

$C = \dfrac{3}{5} + \dfrac{2R}{7} - \dfrac{5}{12} \cdot \left(\dfrac{R}{8}\right)^2$. What is the maximum total amount of crops rounded to the nearest thousand tons

that a 20,000 m^2 field will yield in a particular year when the average daily amount of rain is 12 m^3?

(A) 61,000　　　(B) 62,000　　　(C) 63,000　　　(D) 64,000　　　(E) 65,000

49 A road with a grade of n% means that the height of a car driving on that road changes by n units for every change of 100 units in horizontal distance. According to the table given to the right, a car that travels a distance of 350 miles on road A climbs up an average height of how many miles?

Road	Grade of road (%)	Length of road (mi)
A	6.5	7450
B	5.7	3340
C	9.3	6520
D	4.2	9980
E	8.6	5570

(A) 21.3 mi　　　(B) 22.4 mi　　　(C) 22.7 mi

(D) 23.5 mi　　　(E) 6.55 mi

50. Which of the following gives all possible values of f(x) for the function defined by $f(x) = 1 + \sqrt{25 - x^2}$?

(A) $-25 \le f(x) \le 25$　　(B) $-5 \le f(x) < 5$　　(C) $1 < f(x) \le 6$　　(D) $1 \le f(x) \le 6$　　(E) $f(x) \ge 0$

S T O P

END OF TEST

(Answers on page 202 – Solutions on page 230)

Model Test 14

Test Duration: 60 Minutes

Directions: For each of the following problems, decide which is the **best** of the choices given. If the exact numerical value is not one of the choices, select the choice that best approximates this value. Then fill in the corresponding oval on the answer sheet.

Notes:

- A calculator will be necessary for answering some (but not all) of the questions in this test. For each question you will have to decide whether or not you should use a calculator. The calculator you use must be at least a scientific calculator; programmable calculators and calculators that can display graphs are permitted.

- The only angle measure used on this test is degree measure. Make sure your calculator is in the degree mode.

- Figures that accompany problems in this test are intended to provide information useful in solving the problems. They are drawn as accurately as possible **except** when it is stated in a specific problem that its figure is not drawn to scale.

- All figures lie in a plane unless otherwise indicated.

- Unless otherwise specified, the domain of any function f is assumed to be the set of all real numbers **x** for which **f(x)** is a real number.

Reference Information: The following information is for your reference in answering some of the questions in this test.

- Volume of a right circular cone with radius **r** and height **h**: $V = \frac{1}{3}\pi r^2 h$

- Lateral area of a right circular cone with circumference of the base **c** and slant height **l**: $S = \frac{1}{2}cl$

- Volume of a sphere with radius **r**: $V = \frac{4}{3}\pi r^3$

- Surface area of sphere with radius **r**: $S = 4\pi r^2$

- Volume of a pyramid with base area **B** and height **h**: $V = \frac{1}{3}Bh$

1. $[3 - 10(3 - 10)^{-1}]^{-1} = ?$

(A) 0.0225   (B) 0.0226   (C) 0.325   (D) 0.225   (E) 0.226

2. If $2x^2 + 3 = 7$ then $(2x^4 + 6)/(2x^4 + 5) = ?$

(A) 13/14   (B) 14/13   (C) 9/10   (D) 10/9

(E) None of the above

GO ON TO THE NEXT PAGE ▶▶▶

Model Test 14

3. A factor of $4x^2 + 2x - y^2 - y$ is

(A) $2x + 1$ (B) $2x + y$ (C) $2x + y + 1$ (D) $y + 1$ (E) $2x - y + 1$

4. What is the greatest prime factor of the sum of all prime numbers less than 20?

(A) 2   (B) 3   (C) 7   (D) 11   (E) 13

5. Which of the following is the solution set of the inequality given by $9 - x^2 \geq x - 3$?

(A) $-4 \leq x \leq -3$ (B) $-3 < x < 4$ (C) $-3 \leq x \leq 4$ (D) $-4 < x < 3$ (E) $-4 \leq x \leq 3$

6. What is the sum of all distinct integers between 49 and 499 that give the remainder of 3 when divided by 6?

(A) 19980   (B) 20475   (C) 20481   (D) 40950   (E) 40956

7. In two electronic stores A and B the profit margins are 9.75% and 8.50% respectively. If the price of a certain laptop differs by $15 in the two stores then what is the wholesale price of the laptop assuming that both stores work with the same distributor?

(A) 1100   (B) 1200   (C) 1300   (D) 1400   (E) 1500

8. If x, y and z are the measures of the angles of a triangle and $x = y - z$ then the triangle is

(A) acute   (B) isosceles   (C) obtuse   (D) right   (E) equilateral

GO ON TO THE NEXT PAGE ▶▶▶

Model Test 14

9. Which of the following functions satisfy the relation $f(-x) = -f(x)$

(A) $f(x) = x^4 - 3x^2 + 5$ (B) $f(x) = x^4 + x^3$ (C) $f(x) = |x|$

(D) $f(x) = 2x^3 + x - 5$ (E) $f(x) = -7x^3 + x$

10. In figure 1, which quadrants contain pairs (x, y) that satisfy the condition $\frac{y}{x} > 1$?

(A) I only (B) I and II only (C) I and III only (D) II and IV only (E) I, II and III only

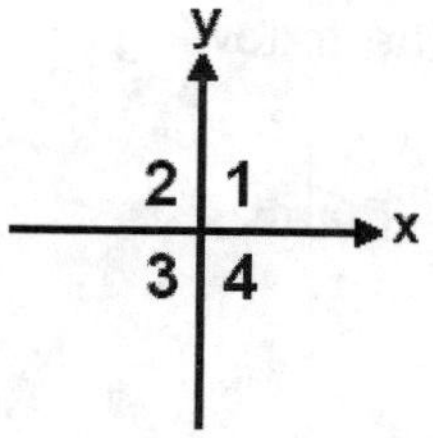

Figure 1

11. A square is inscribed in a circle of radius 2. What is the length of one side of the square?

(A) 1 (B) 2 (C) 4 (D) $\sqrt{2}$ (E) $2\sqrt{2}$

12. In a rectangular box 400 cubes of sugar can fit. How many cubes can fit in a box three times as wide, three times as long, and three times as high?

(A) 1200 (B) 1800 (C) 3600 (D) 7200 (E) 10800

13. Merve earns 30$ an hour, of which 2.55% is cut for taxes. How many cents per hour of her income goes to taxes?

(A) 0.765 (B) 7.65 (C) 76.5 (D) 2.65 (E) 26.5

GO ON TO THE NEXT PAGE ▶▶▶

Model Test 14

14. In how many ways can 5 people sit at a round table if Bilal and Hatice must sit next to each other?

(A) 12   (B) 18   (C) 24   (D) 48   (E) 120

15. If the line shown in figure 2 is the graph of $y = ax - b$ then which of the following must be true?

  (A) $ab > 1$ (B) $0 < ab < 1$ (C) $ab = 1$

  (D) $ab < -1$ (E) $-1 < ab < 0$

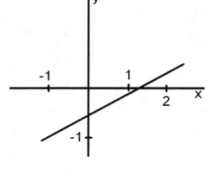

Figure 2

16. Zeynep and Emine went shopping. They spent half of what they had, plus \$2 at the first store. At the second store, they spent half of what was left, plus \$5. At the third store, they spent half of what was left. They spent the remaining \$5 on hot chocolates. How much did they start with?

(A) \$100   (B) \$88   (C) \$64   (D) \$45   (E) It cannot be determined from the given information.

17. Which of the following implies that a given quadrilateral is a rectangle?

    I. If it has two pairs of opposite sides that are parallel and equal in length.

    II. If its diagonals bisect each other and are equal in length.

    III. If three of its exterior angles measure 90° each.

(A) I only   (B) II only   (C) III only   (D) II and III   (E) I, II and III

18. In a parallelogram with sides of length 5 and 12, one of the interior angles is obtuse. If the angle that the longer diagonal makes with the shorter side is θ, then which of the following is incorrect?

(A) $\sin\theta < 12/13$ (B) $\sin^2(\theta)+\cos^2(\theta)=1$ (C) $\tan\theta > 12/5$

(D) $\cos\theta > 5/13$ (E) $1+\tan^2\theta=1/\cos^2\theta$

GO ON TO THE NEXT PAGE ▶▶▶

Model Test 14

19. Two junior skiers Incilay and Ozge, and two senior skiers, Simin and Burcu, go on a vacation at Kartalkaya. However in the hotel that they must stay there are only three rooms available, and each senior must be alone in a room. The room assignment of all four skiers can be determined from the assignment of which of the following pairs?

    I. Incilay and Ozge

    II. Ozge and Simin

    III. Simin and Burcu

(A) I only   (B) II only   (C) III only   (D) II and III   (E) I, II and III

20. If the radius of the circle given in figure is 5, what is the equation of the circle?

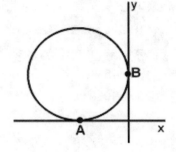

Figure 3

(A) $(x - 5)^2 + (y - 5)^2 = 25$ (B) $(x + 5)^2 + (y - 5)^2 = 25$

(C) $(x - 5)^2 + (y + 5)^2 = 25$ (D) $(x + 5)^2 + (y - 5)^2 = 5$

(E) $(x - 5)^2 + (y + 5)^2 = 5$

21. The probability that Melek hits a certain target is $\frac{2}{5}$ and, independently, the probability that Mukadder hits it is $\frac{4}{7}$. What is the probability that Melek hits the target and Mukadder misses it?

(A) $\frac{3}{5} \cdot \frac{4}{7}$ (B) $\frac{2}{5} \cdot \frac{4}{7}$ (C) $\frac{2}{5} + \frac{3}{7}$ (D) $\frac{2}{5} \cdot \frac{3}{7}$ (E) $\frac{2}{5} + \frac{3}{7} - \frac{2}{5} \cdot \frac{3}{7}$

22. The sequence given by 0, 2, 6, 14, 30, 62... follows a simple pattern. What is the next number in the sequence?

(A) 116   (B) 126   (C) 136   (D) 146   (E) 156

GO ON TO THE NEXT PAGE ▶▶▶

Model Test 14

23. How many lines can partition a square into two congruent regions?

(A) no line   (B) 1 line   (C) 2 lines   (D) 4 lines   (E) more than 4 lines

24. Which of the following intervals is a superset of all other intervals?

(A) $1 < x < 7$ (B) $2 \le x < 6$ (C) $3 \le x \le 4$ (D) $1 \le x < 7$ (E) $1 \le x \le 5$

25. In figure 4 given below, BD and CD are the bisectors of the angles $\angle$ QBC and $\angle$ PCB respectively. What is the relation between the measures of angles x and y?

(A) $y = 45° - x$ (B) $x = 45° + y$ (C) $y = 2x/3$

(D) $x = 90° - 0.5y$ (E) $y + 0.5x = 90°$

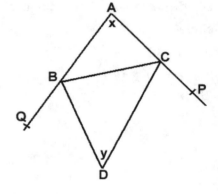

Figure 4
Figure not drawn to scale

26. What is the area of a triangle with sides of length 26, 28, and 30?

(A) 339   (B) 168   (C) 84   (D) 336   (E) 166

27. It is given that $ma + nb = 7$ and $na + mb = 5$. If $m - n = 0.4$ then $a^2 + b^2 - 2ab = ?$

(A) 0   (B) 2   (C) 4   (D) 5   (E) 25

GO ON TO THE NEXT PAGE ▶▶▶

28. In figure 5, EB and MR are parallel line segments. What is the area in square yards of triangle BUR if the lengths of EU and MR are 32 and 27 feet respectively?

(A) 48   (B) 96   (C) 432   (D) 864

(E) It cannot be determined from information given.

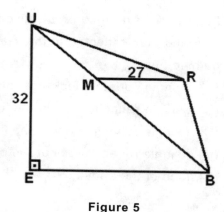

Figure 5

Figure not drawn to scale

29. If θ is a positive acute angle in the equation given by $2\sin(3\theta) - \cos^2(2\theta) = \sin^2(2\theta)$, then the least possible value of θ is

(A) 10°   (B) 20°   (C) 30°   (D) 40°   (E) 50°

30. In the rectangular box given in figure 6, how many different paths of total length 18 from A to B are there along the edges?

(A) 4   (B) 5   (C) 6   (D) 7   (E) 8

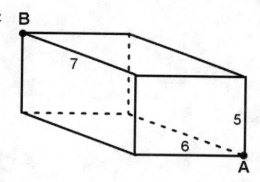

Figure 6

Figure not drawn to scale

31. Let x_1 and x_2 be the zeros of a quadratic polynomial P(x). If $x_1 + x_2 = -2$ and $x_1 \cdot x_2 = 3$, which of the following could be the polynomial P(x)?

(A) $2x^2 - x + 3$ (B) $3x^2 + x + 2$ (C) $2x^2 + 4x + 6$ (D) $-2x^2 - 4x + 3$ (E) $x^2 - 3x + 2$

32. The parametric equations given by $x = -t^2$ and $y = t^4 - 1$ represent

(A) a line   (B) a parabola   (C) portion of a line   (D) portion of a parabola   (E) portion of a hyperbola

GO ON TO THE NEXT PAGE ▶▶▶

33. The right triangle ΔABC given in figure 7 is first reflected in the y axis and then reflected in the x axis. The resulting figure can also be obtained by

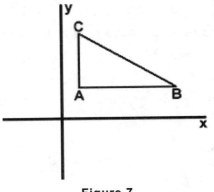

(A) a single counter clockwise rotation about the origin

(B) a single reflection about the line y = -x

(C) a single reflection about the line y = x

(D) a single reflection about the origin

(E) none of the above

Figure 7

34. If in figure 8 P is a variable point on BC then AP + PD can be at most

(A) 15   (B) 16.7   (C) 17.5   (D) 17.6   (E) 21

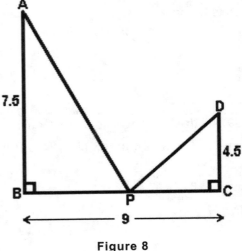

Figure 8

35. The solution set of f(x) = A is {2, -3}. What is the solution set of f(x − 3) = A?

(A) {2, -3}   (B) {-5, -6}   (C) {5, 0}   (D) {2, 3}   (E) {-2, -3}

36. The line y=x+1 is reflected across the point A(1, -1) to get the line l. What is the equation of line l?

(A) y = x+5   (B) y = -x+5   (C) y = x-0.5   (D) y = x-5   (E) y = -x-5

GO ON TO THE NEXT PAGE ▶▶▶

Model Test 14

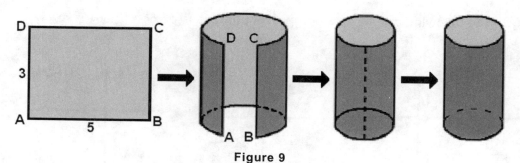

Figure 9

Figure not drawn to scale

37. The rectangular piece of paper ABCD given in figure 9 is rolled into a cylindrical tube without overlap and covered on top and bottom surfaces. If the dimensions of the rectangle ABCD are in inches, what is the volume of the resulting cylinder in cubic inches?

(A) 5.96   (B) 5.97   (C) 15   (D) 185   (E) 581.37

38. The "reach" of a point in the xy coordinate plane is defined as |x| + |y| where (x,y) are the coordinates of the point. Which of the following points has the same reach as $\left(\dfrac{-3}{2},\dfrac{1}{2}\right)$?

     I. (-1, 1)   II. (0, -2)   III. (0.5, 0.5)

(A) I only   (B) II only   (C) I and II only   (D) II and III only   (E) I, II and III

39. The graph of f(x)= $-x^2$ is translated 2 units left and 1 unit up to represent g(x). What is the value of g(-1.2)?

(A) 0.36   (B) -0.36   (C) 9.24   (D) -9.24   (E) 1.64

40. If f(x) = 6x − 3 then what is the slope of the line given by y = f(-2x+1)?

(A) -12   (B) -6   (C) -2   (D) 6   (E) 12

GO ON TO THE NEXT PAGE ▶▶▶

Model Test 14

41. How many of the following graphs do not represent functions of x?

I.	II.	III.	IV.	V.

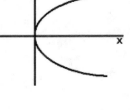 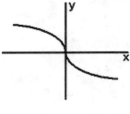 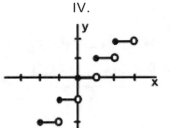 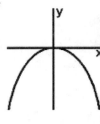

(A) 1   (B) 2   (C) 3   (D) 4   (E) 5

42. Which of the following is not correct?

(A) Reflection of the point (x, y) in the x – axis gives the point (x, -y).

(B) Reflection of the point (x, y) in the y – axis gives the point (-y, x).

(C) Reflection of the point (x, y) in the origin gives the point (-x, -y).

(D) Reflection of the point (x, y) in the line y = x gives the point (y, x).

(E) Reflection of the point (x, y) in the line y = - x gives the point (-y, -x).

43. Which of the following is an example of horizontal stretch?

(A)	(B)	(C)	(D)	(E)

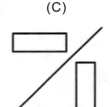

44. If $P(x)=ax^2+bx+c$ for all real numbers x and if $P(0) = 3$ and $P(-1) = 5$, then $a - b = ?$

(A) -2   (B) -1   (C) 0   (D) 1   (E) 2

GO ON TO THE NEXT PAGE ►►►

Model Test 14

45. What is the radius of the circle given by $x^2 - 6x + y^2 + 4y = 1$?

(A) 3   (B) $\sqrt{10}$ (C) $\sqrt{12}$ (D) $\sqrt{14}$ (E) 4

46. From an airplane flying at a height of 1972 feet over the ocean, the angle of depression of an island is 31.5°. Approximately how many miles must the airplane travel horizontally to be directly over the island?

(A) 3000   (B) 3100   (C) 3200   (D) 3300   (E) 3400

47. It is given that $f(x)$ is a linear function having a slope of 3 and passing through the point (5,17); $f(0)$=?

(A) – 2   (B) 2   (C) 3   (D) 11   (E) 17

48. How many integers are there in the solution set of the following inequality given by $|2x - 8| > 3$?

(A) None   (B) 3   (C) 4   (D) 5   (E) More than 5

49. Figure 10 shows the side views of two identical ladders that lean against two walls directly opposite each other each being 25 feet long. Ladder MN is stationary while ladder AB is sliding down. How many feet must the A end of ladder AB slide down so that tanα and tanβ will be equal?

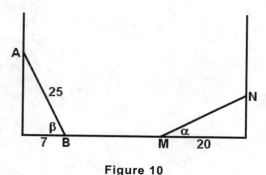

(A) 9   (B) 11   (C) 13   (D) 15

(E) None of the above

Figure 10

Figure not drawn to scale

GO ON TO THE NEXT PAGE ▶▶▶

Model Test 14

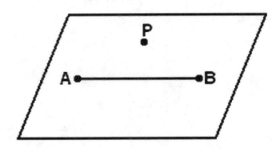

Figure 11

50. Point P and segment AB are in the same plane and P is a variable point above segment AB as indicated in figure 11. If PAB is an obtuse triangle where the measure of P is always 120° then the locus of all such points P is

(A) a semicircle whose diameter is AB.

(B) part of a circle whose diameter is AB.

(C) a circle whose diameter is greater than AB.

(D) the minor arc AB of a circle whose diameter is greater than AB.

(E) the major arc AB of a circle whose diameter is greater than AB.

S T O P

END OF TEST

(Answers on page 202 – Solutions on page 232)

Model Test 15

Test Duration: 60 Minutes

Directions: For each of the following problems, decide which is the **best** of the choices given. If the exact numerical value is not one of the choices, select the choice that best approximates this value. Then fill in the corresponding oval on the answer sheet.

Notes:

- A calculator will be necessary for answering some (but not all) of the questions in this test. For each question you will have to decide whether or not you should use a calculator. The calculator you use must be at least a scientific calculator; programmable calculators and calculators that can display graphs are permitted.

- The only angle measure used on this test is degree measure. Make sure your calculator is in the degree mode.

- Figures that accompany problems in this test are intended to provide information useful in solving the problems. They are drawn as accurately as possible **except** when it is stated in a specific problem that its figure is not drawn to scale.

- All figures lie in a plane unless otherwise indicated.

- Unless otherwise specified, the domain of any function f is assumed to be the set of all real numbers **x** for which **f(x)** is a real number.

Reference Information: The following information is for your reference in answering some of the questions in this test.

- Volume of a right circular cone with radius **r** and height **h**: $V = \frac{1}{3}\pi r^2 h$

- Lateral area of a right circular cone with circumference of the base **c** and slant height **l**: $S = \frac{1}{2}cl$

- Volume of a sphere with radius **r**: $V = \frac{4}{3}\pi r^3$

- Surface area of sphere with radius **r**: $S = 4\pi r^2$

- Volume of a pyramid with base area **B** and height **h**: $V = \frac{1}{3}Bh$

1. Which of the following illustrates a distributive principle?

(A) $(7 + 3) + 5 = 7 + (3 + 5)$

(B) $(7 \cdot 3) \cdot 5 = 7 \cdot (3 \cdot 5)$

(C) $7 + 3 = 3 + 7$

(D) $7 \cdot (3 + 5) = 7 \cdot 3 + 7 \cdot 5$

(E) $7 \cdot 3 + 5 = 3 \cdot 7 + 5$

GO ON TO THE NEXT PAGE ▶▶▶

Model Test 15

2. In the figure 1, AB = 10, BC = 2, and D is 4 times as far from A as from C. What is CD?

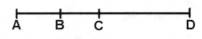

Figure 1
Figure not drawn to scale

  (A) 2   (B) 4   (C) 6   (D) 10   (E) 12

3. If n is a positive integer, which of the following is always even?

     I. $n^2 + 3n + 1$

     II. $2n^2 - 4n$

     III. $n^2 - n$

(A) I only   (B) II only   (C) III only   (D) II and III only   (E) I, II and III

4. Let A be the set of all numbers a, such that $-3 < a < 7$. Let B be the set of all numbers b such that $5 \le b < 10$. The intersection, C, of A and B is the set of all numbers c such that

(A) $-3 < c \le 5$ (B) $-3 < c < 7$ (C) $5 \le c < 7$ (D) $5 \le c < 10$ (E) $-3 < c < 10$

5. If $a > 1$ and $a^r = 2.4$, then $a^{-2r} =$?

(A) -4.8 (B) -5.76 (C) $-\dfrac{1}{5.76}$ (D) $\dfrac{1}{5.76}$ (E) $\dfrac{1}{4.8}$

6. The inequality $(x - 3)^2 \cdot (x+5) > 0$ is valid if and only if which of the following conditions is satisfied?

(A) $x < -5$ (B) $x > -5$ or $x \neq 3$ (C) $-5 < x < 3$ (D) $x < -5$ and $x \neq 3$ (E) $x > -5$ and $x \neq 3$

GO ON TO THE NEXT PAGE ▶▶▶

Model Test 15

7. A computer is programmed to subtract 5 from M, multiply the result by 4, add 20, and divide the final quantity by 4. The answer given by the computer will be

(A) M (B) M − 5 (C) 4·M (D) M+20 (E) 4·M+20

8. What is the value of $\dfrac{2x-8}{x-8} \cdot \dfrac{x^2-5x-24}{x^2-16}$ where defined?

(A) 2 (B) x − 4 (C) x + 8 (D) $\dfrac{2x+6}{x+4}$ (E) $\dfrac{2x^2-10x-48}{x^2-16}$

9. If $S+e = \dfrac{S-e}{t}$, e =

(A) $\dfrac{S \cdot (1+t)}{1-t}$ (B) $\dfrac{S \cdot (1-t)}{1+t}$ (C) $\dfrac{1-t}{S \cdot (1+t)}$ (D) $\dfrac{1+t}{S \cdot (1-t)}$ (E) $\dfrac{1+t}{1-t}$

10. If $f^{-1}(x) = \dfrac{x}{4} - \dfrac{1}{3}$ and f(x) = ?

(A) $\dfrac{12x-4}{3}$ (B) $\dfrac{4}{3}(3x+1)$ (C) $\dfrac{4}{3}(-3x+1)$ (D) $\dfrac{3}{4}(4x+1)$ (E) $\dfrac{3}{4}(3x-1)$

11. If A varies inversely as $\dfrac{1}{B}$ and A = 8 when B = 4; what is A^2 when B = 5?

(A) 10 (B) 16 (C) 25 (D) 64 (E) 100

GO ON TO THE NEXT PAGE ▶▶▶

Model Test 15

12. How many degrees is the angle between the hands of a clock at 7:20?

(A) 90°   (B) 96°   (C) 97°   (D) 98°   (E) 100°

13. According to the data given in figure 2, what is the radius of the circle passing through the vertices of the shaded triangle?

(A) 13.2   (B) 16.3   (C) 19.4   (D) 26.5   (E) 32.6

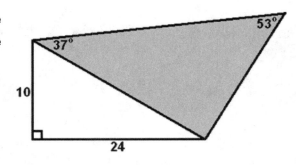

Figure 2

Figure not drawn to scale

14. Legs of a right triangle have lengths that are in the ratio of 2:5. If the area of the triangle is 20, what is the length of its hypotenuse?

(A) 10   (B) $\sqrt{21}$ (C) $\sqrt{29}$ (D) $2\sqrt{21}$ (E) $2\sqrt{29}$

15. The two rectangles given in figure 3 are symmetrical with respect to line m. If the area of the shaded square in the left hand rectangle is 1 square feet, what is the area in square inches of the shaded area in the right hand rectangle?

(A) 1   (B) 12   (C) 120   (D) 144

(E) It cannot be determined from the information given

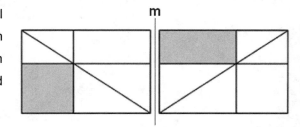

Figure 3

GO ON TO THE NEXT PAGE ▶▶▶

Model Test 15

16. Point A and circle O are in the same plane. Of all points on circle O, the one that is closest to point A is another point B and length of AB is 8 inches. If the circle has a radius of 8 inches, what is the length, in feet, of the tangent from point A to the circle?

(A) 1.39   (B) 13.9   (C) 1.15   (D) 11.5   (E) 8

17. In the formula $p = \dfrac{r}{A}$, if $r = 8 \times 10^8$ and $p = 4 \times 10^{-4}$, A = ?

(A) 2×10^{-11} (B) 5×10^{11} (C) 2×10^{12} (D) 5×10^{12} (E) 5×10^{12}

18. What is the approximate slope of the line $\sqrt{5}x + 5y - 2\sqrt{5} = 0$?

(A) -2.45   (B) -2.24   (C) -1.24   (D) -0.44   (E) -0.45

19. How many numbers in the set $\{-5, 0, 4, 12, 15\}$ satisfy the condition $|x - 4| \geq 8$?

(A) none   (B) one   (C) two   (D) three   (E) four

20. For what value of x does the function $f(x) = -x^2 - \sqrt{3}x + 4$ become maximum?

(A) -3.04   (B) -0.87   (C) 0   (D) 0.87   (E) 2.14

GO ON TO THE NEXT PAGE ▶▶▶

Model Test 15

21. In $\triangle ABC$, the measure of $\angle B$ is 45° and the measure of $\angle A$ is a. If $|AB|$ is longer than $|AC|$, then

(A) 0° < a < 45°   (B) 0° < a < 90°   (C) 45° < a < 90°   (D) 45° < a < 135°   (E) 90° < a < 135°

22. Four parallel lines are cut by three nonparallel lines. What is the maximum number of points of intersection points that can be obtained?

(A) 9   (B) 10   (C) 12   (D) 13   (E) 15

23. What is the area in square inches of the largest rectangle that can fit into the rhombus given in figure 4 if each side of the rhombus has a length of 20 inches and one of the interior angles of the rhombus measures 50° given that all vertices of the rectangle are on the rhombus?

(A) 131   (B) 132   (C) 133   (D) 265   (E) 266

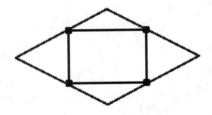

Figure 4
Figure not drawn to scale

24. If $f(x) = (c - 2) \cdot x + c + 1$ is the same for all x, then f(2007)=?

(A) 3   (B) 1   (C) 2008   (D) 4013

(E) It cannot be determined from the information given.

25. If (x, y) is a point on the function $f(x) = x^2 + x - 3$, for what value(s) of x will y be three times x?

(A) no value   (B) -1 only   (C) 2 only   (D) 3 only   (E) -1 and 3

GO ON TO THE NEXT PAGE ▶▶▶

Model Test 15

26. In the semicircle given in figure 5, O is the center of the semicircle, $|AC| = 3$ inches and, $|BC| = 4$ inches. If C is a movable point on arc AB, then the greatest possible area in square inches of ΔACB is

(A) 50   (B) 25   (C) 12.5   (D) 6.25

(E) There is not enough information to solve this problem.

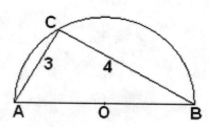

Figure 5
Figure not drawn to scale

27. A triangle with vertices (-1, -1), (3, -5), and (3, -1) belongs to which of the following classes?

    I. Right Triangles   II. Isosceles Triangles

    III. Scalene Triangles   IV. Equilateral Triangles

(A) I only   (B) II only   (C) III only   (D) IV only   (E) I and II only

28. What is the slope of the perpendicular bisector of segment MT that joins the points M(2,-5) and T(-1,3)?

(A) 0.37   (B) 0.38   (C) -3/8   (D) 8/3   (E) -8/3

29. The graph given in figure 6 corresponds to which of the following?

(A) $y = |x - 2|$ (B) $y = |x + 2|$ (C) $y = x - 2$ (D) $y = x$ (E) $y = x + 2$

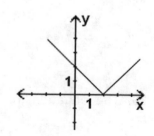

Figure 6

30. If the graph of the equation $2x - y + 6 - 2k = 0$ passes through the origin, the value of k is

(A) 3   (B) 1   (C) 0   (D) -1   (E) -3

GO ON TO THE NEXT PAGE ▶▶▶

Model Test 15

31. If x = - 64, the value of $\dfrac{1}{4}x^{\frac{2}{3}} + x^{\frac{1}{3}}$ is

(A) 0   (B) 2   (C) 4   (D) 8   (E) 16

32. If $3^{2x-6} = \left(\dfrac{1}{9}\right)^x$ then x = ?

(A) $\dfrac{1}{3}$ (B) $\dfrac{1}{9}$ (C) $\dfrac{3}{2}$ (D) $\dfrac{2}{9}$ (E) 6

33. The curve that $x^2 + y^2 = 25$ represents which of the following?

(A) a circle   (B) an ellipse   (C) a hyperbola   (D) a parabola   (E) a straight line

34. Which of the following does the fraction $\dfrac{\sqrt{18} + \sqrt{12}}{\sqrt{6}}$ equal to?

(A) $3 + \sqrt{12}$ (B) $\sqrt{2} + \sqrt{3}$ (C) 5   (D) $2\sqrt{3}$ (E) $3\sqrt{2}$

35. The equation that expresses the relationship between x and y in the table is

x	1	3	4	8
y	0	4	6	14

(A) y - 2x – 2 = 0   (B) y = x – 1   (C) y – 2x + 2 = 0

(D) y = 3x – 3   (E) y + 2x + 2 = 0

GO ON TO THE NEXT PAGE ▶ ▶ ▶

Model Test 15

36. For what values of A does the equation $x^2 - Ax + A = 0$ have no real roots?

(A) $0 \le A \le 4$ (B) $0 < A < 4$ (C) $-4 \le A \le 0$ (D) $A \le 0$ or $A \ge 4$ (E) $A \le -4$

37. If $f(x)$ is a linear function such that $f(x - 1) + f(x + 1) = 6x - 10$ then $f(1) = ?$

(A) -8   (B) -5   (C) -3   (D) -2   (E) 2

38. Which of the following expresses the infinite decimal 0.2161616... as a common fraction?

(A) $\dfrac{216}{990}$ (B) $\dfrac{214}{999}$ (C) $\dfrac{214}{99}$ (D) $\dfrac{214}{990}$ (E) $\dfrac{216}{1000}$

39. In figure 7, ⌊PA and ⌊PB are tangent to circle O. If angle P measures 80°, how many degrees are in the major arc AB?

(A) 60   (B) 180   (C) 260   (D) 100   (E) 120

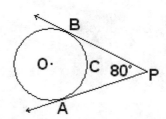

Figure 7

Figure not drawn to scale

40. A cubic foot of water is poured into a rectangular tank whose base dimensions are 22 inches by 10 inches. To what height in inches does the water rise if the height of the tank is greater than 14 inches?

(A) 0.13   (B) 0.65   (C) 1.53   (D) 6.11   (E) 7.85

GO ON TO THE NEXT PAGE ▶▶▶

Model Test 15

41. A bus travels a distance of t miles at 60 mph and returns at 80 mph. What is the average rate of the bus for the round trip?

(A) 68.57   (B) 70   (C) 71.42   (D) 72   (E) 73

42. If $y = \sqrt{11}x^2 - \sqrt{3}x - \sqrt{2}$, what is the approximate product of the roots?

(A) 0.52   (B) -0.43   (C) -2.35   (D) -1.91   (E) -0.82

43. If x is a nonzero real number, which of the following lead to the fact that x is positive?

     I. $|x| = -x$

     II. $|x| \leq x$

     III. $|x| > x$

(A) I only   (B) II only   (C) III only   (D) II and III only   (E) I, II and III

44. Each interior angle of a regular polygon measures 160°. How many diagonals does the polygon have?

(A) 18   (B) 20   (C) 40   (D) 80   (E) 135

45. The solution set of the equation $x - 4\sqrt{x} - 5 = 0$ is

(A) {1}   (B) {1, 5}   (C) {-5}   (D) {5}   (E) {25}

GO ON TO THE NEXT PAGE ▶▶▶

Model Test 15

46. The graphs of the equations $x^2 + y^2 = 16$ and $y = x^2 - 4$ intersect at how many points?

(A) 0   (B) 1   (C) 2   (D) 3   (E) 4

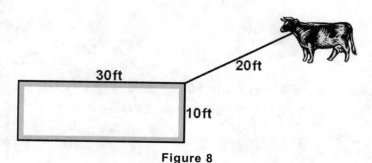

Figure 8
Figure not drawn to scale

47. What is the bull's total gazing area rounded to the nearest 10 square feet if the bull is attached to the corner of a 30 ft by 10 ft rectangular wall as shown in figure 8 above?

(A) 320   (B) 330   (C) 1000   (D) 1010   (E) 1020

48. The graph of f(x) that is symmetric with respect to the line x = 1 is given in figure 9 and g(x) is defined in terms of f(x) as follows: g(x) = f(x) and x ≤ 1. Which of the following functions equals $g^{-1}(x)$?

(A) $2x - 2$ (B) $-2x + 2$ (C) $1 - 2x$ (D) $\dfrac{2+x}{2}$ (E) $\dfrac{2-x}{2}$

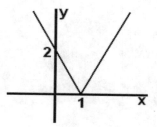

Figure 9

GO ON TO THE NEXT PAGE ▶▶▶

Model Test 15

49. Filiz can complete a job in h/3 hours alone and Melissa can complete it in 2h/5 hours alone. If they both work together, how many hours in terms of h will it take them to complete three such jobs?

(A) $\dfrac{26h}{15}$ (B) $\dfrac{15}{26h}$ (C) $\dfrac{6h}{11}$ (D) $\dfrac{2h}{11}$ (E) $\dfrac{11h}{6}$

50. How many distinct squares are there in figure 10?

  (A) 25   (B) 40   (C) 50   (D) 55   (E) More than 55

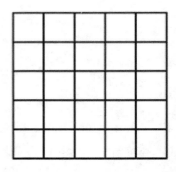

Figure 10

S T O P

END OF TEST

(Answers on page 202 – Solutions on page 234)

ANSWERS TO MODEL TESTS

Model Test 1 Answers

1	C	11	A	21	C	31	C	41	B
2	D	12	C	22	A	32	D	42	D
3	C	13	C	23	E	33	D	43	B
4	E	14	E	24	A	34	C	44	B
5	E	15	A	25	D	35	C	45	B
6	C	16	D	26	B	36	C	46	D
7	A	17	C	27	A	37	A	47	C
8	A	18	D	28	D	38	B	48	B
9	B	19	E	29	B	39	C	49	B
10	A	20	D	30	E	40	E	50	A

Model Test 2 Answers

1	A	11	C	21	B	31	E	41	D
2	E	12	D	22	C	32	B	42	E
3	D	13	B	23	B	33	C	43	D
4	C	14	B	24	B	34	E	44	D
5	B	15	C	25	D	35	D	45	C
6	D	16	C	26	A	36	C	46	C
7	E	17	B	27	B	37	E	47	C
8	B	18	E	28	C	38	C	48	C
9	D	19	C	29	C	39	B	49	B
10	D	20	C	30	C	40	B	50	C

Model Test 3 Answers

1	D	11	C	21	E	31	D	41	B
2	B	12	B	22	B	32	D	42	C
3	B	13	D	23	A	33	E	43	E
4	C	14	E	24	B	34	B	44	B
5	A	15	D	25	A	35	C	45	B
6	D	16	C	26	C	36	E	46	A
7	A	17	A	27	D	37	E	47	A
8	B	18	C	28	A	38	B	48	B
9	D	19	A	29	D	39	D	49	B
10	E	20	B	30	D	40	A	50	D

Model Test 4 Answers

1	E	11	C	21	D	31	E	41	E
2	A	12	A	22	A	32	D	42	C
3	E	13	D	23	D	33	E	43	D
4	A	14	C	24	B	34	C	44	D
5	B	15	E	25	C	35	C	45	A
6	E	16	A	26	C	36	B	46	D
7	B	17	E	27	E	37	B	47	B
8	D	18	B	28	C	38	A	48	B
9	C	19	C	29	C	39	A	49	B
10	B	20	B	30	B	40	B	50	D

Model Test 5 Answers

1	C	11	A	21	D	31	E	41	D
2	E	12	A	22	E	32	E	42	C
3	E	13	A	23	D	33	C	43	A
4	D	14	A	24	A	34	E	44	E
5	D	15	C	25	B	35	A	45	D
6	E	16	B	26	C	36	E	46	C
7	D	17	D	27	C	37	C	47	D
8	B	18	D	28	D	38	D	48	C
9	D	19	E	29	A	39	E	49	B
10	E	20	E	30	C	40	D	50	C

Model Test 6 Answers

1	C	11	C	21	A	31	D	41	B
2	A	12	E	22	D	32	E	42	B
3	D	13	A	23	A	33	C	43	B
4	E	14	E	24	D	34	C	44	D
5	E	15	D	25	D	35	D	45	E
6	E	16	E	26	D	36	E	46	A
7	D	17	E	27	D	37	C	47	C
8	D	18	C	28	C	38	A	48	B
9	A	19	B	29	E	39	D	49	D
10	C	20	D	30	B	40	C	50	C

Model Test 7 Answers

1	C	11	C	21	D	31	C	41	D
2	B	12	E	22	E	32	C	42	C
3	E	13	B	23	D	33	A	43	E
4	E	14	B	24	E	34	B	44	E
5	C	15	C	25	C	35	C	45	E
6	D	16	D	26	C	36	E	46	C
7	A	17	C	27	B	37	E	47	E
8	D	18	E	28	E	38	C	48	D
9	C	19	C	29	E	39	C	49	E
10	D	20	B	30	C	40	B	50	E

Model Test 8 Answers

1	B	11	A	21	B	31	E	41	D
2	D	12	C	22	A	32	C	42	D
3	B	13	C	23	A	33	C	43	B
4	C	14	E	24	B	34	A	44	C
5	D	15	C	25	C	35	E	45	C
6	E	16	D	26	C	36	B	46	D
7	D	17	C	27	B	37	B	47	C
8	C	18	D	28	B	38	B	48	C
9	A	19	A	29	D	39	A	49	B
10	C	20	D	30	E	40	B	50	C

ANSWERS TO MODEL TESTS

Model Test 9 Answers

#		#		#		#		#	
1	C	11	C	21	C	31	C	41	A
2	B	12	D	22	A	32	E	42	C
3	D	13	A	23	D	33	D	43	D
4	A	14	D	24	A	34	B	44	C
5	B	15	C	25	C	35	D	45	D
6	A	16	D	26	E	36	D	46	E
7	C	17	D	27	B	37	B	47	C
8	E	18	A	28	D	38	B	48	E
9	E	19	E	29	C	39	D	49	B
10	C	20	B	30	C	40	B	50	B

Model Test 10 Answers

#		#		#		#		#	
1	B	11	D	21	B	31	B	41	E
2	B	12	E	22	B	32	C	42	E
3	B	13	E	23	D	33	E	43	D
4	C	14	B	24	A	34	A	44	C
5	B	15	C	25	C	35	D	45	E
6	E	16	E	26	A	36	B	46	A
7	C	17	B	27	C	37	B	47	C
8	E	18	A	28	D	38	A	48	B
9	D	19	B	29	C	39	C	49	C
10	A	20	D	30	E	40	D	50	B

Model Test 11 Answers

#		#		#		#		#	
1	E	11	D	21	C	31	D	41	C
2	C	12	B	22	C	32	E	42	D
3	D	13	A	23	C	33	B	43	D
4	D	14	C	24	A	34	C	44	C
5	C	15	D	25	D	35	B	45	A
6	B	16	D	26	B	36	B	46	B
7	B	17	C	27	C	37	C	47	A
8	E	18	D	28	C	38	B	48	A
9	D	19	B	29	B	39	C	49	E
10	A	20	C	30	E	40	D	50	B

Model Test 12 Answers

#		#		#		#		#	
1	C	11	D	21	E	31	B	41	C
2	C	12	E	22	A	32	B	42	B
3	C	13	B	23	A	33	D	43	E
4	C	14	A	24	C	34	E	44	A
5	D	15	A	25	D	35	C	45	B
6	A	16	B	26	E	36	B	46	C
7	B	17	A	27	B	37	B	47	D
8	B	18	D	28	C	38	E	48	E
9	E	19	D	29	D	39	E	49	C
10	B	20	A	30	E	40	C	50	E

Model Test 13 Answers

#		#		#		#		#	
1	D	11	D	21	A	31	B	41	A
2	E	12	E	22	B	32	A	42	B
3	D	13	A	23	B	33	C	43	B
4	A	14	A	24	B	34	C	44	E
5	D	15	D	25	B	35	A	45	C
6	E	16	C	26	C	36	B	46	C
7	C	17	B	27	E	37	C	47	A
8	D	18	C	28	B	38	A	48	B
9	A	19	C	29	C	39	A	49	C
10	C	20	C	30	D	40	E	50	D

Model Test 14 Answers

#		#		#		#		#	
1	E	11	E	21	D	31	C	41	B
2	B	12	E	22	B	32	D	42	B
3	C	13	C	23	E	33	D	43	B
4	D	14	A	24	D	34	D	44	E
5	E	15	B	25	E	35	C	45	D
6	B	16	C	26	D	36	D	46	C
7	B	17	D	27	E	37	B	47	B
8	D	18	C	28	A	38	C	48	E
9	E	19	D	29	A	39	A	49	A
10	C	20	B	30	C	40	A	50	D

Model Test 15 Answers

#		#		#		#		#	
1	D	11	E	21	B	31	A	41	A
2	B	12	E	22	E	32	C	42	B
3	D	13	B	23	C	33	A	43	B
4	C	14	E	24	A	34	B	44	E
5	D	15	D	25	E	35	C	45	E
6	E	16	C	26	D	36	B	46	D
7	A	17	C	27	E	37	D	47	E
8	D	18	E	28	B	38	D	48	E
9	B	19	D	29	A	39	C	49	C
10	B	20	B	30	A	40	E	50	D

Scaled Score Conversion Table
Mathematics Level 1 Subject Test

Raw Score	Scaled Score	Raw Score	Scaled Score	Raw Score	Scaled Score
50	800	28	590	6	390
49	790	27	580	5	380
48	780	26	570	4	380
47	780	25	560	3	370
46	770	24	550	2	360
45	750	23	540	1	350
44	740	22	530	0	340
43	740	21	520	-1	340
42	730	20	510	-2	330
41	720	19	500	-3	320
40	710	18	490	-4	310
39	710	17	480	-5	300
38	700	16	470	-6	280
37	690	15	460	-7	270
36	680	14	460	-8	260
35	670	13	450	-9	260
34	660	12	440	-10	250
33	650	11	430	-11	240
32	640	10	420		
31	630	9	420		
30	620	8	410		
29	600	7	400		

Blank Page

Model Test 1 – Solutions

1. C

$$\frac{x}{5} + \frac{4x}{5} = 2 \;\rightarrow\; \frac{5x}{5} = 2 \;\rightarrow\; x = 2$$

2. D

$$f(-3) = -3 - 3(-3)^3 = -3 + 81 = 78$$

3. C

$$\left. \begin{array}{l} x = \dfrac{8y}{7} \\[2mm] z = \dfrac{9y}{5} \end{array} \right\} \; \frac{x}{2} = \frac{8}{7}\cdot\frac{5}{9} = \frac{40}{63}$$

4. E

$-(3x-y) = -3x+y$ and rest is false.

5. E

$$\left. \begin{array}{l} x + y = -3 \Rightarrow y = -3 - x \\[2mm] y + z = 2 \Rightarrow y = 2 - z \end{array} \right\} 2 - z = -3 - x \qquad 5 = z - x$$

6. C

m-n can be negative so it is not always a natural number.

7. A

$$5^x \cdot 25^{2x} = 5^x \cdot (5^2)^{2x} = 5^x \cdot 5^{4x} = 5^{5x}$$

8. A

$$\left(\sqrt[3]{p}\right)^2 = 7$$

$$\Rightarrow \sqrt[3]{p} = \mp\sqrt{7} \Rightarrow p = \left(\mp\sqrt{7}\right)^3 = \mp 18.52$$

9. B

$$60° - 30° \cdot \frac{20}{60} = 60° - 10° = 50°$$

10. A

Remaining number is
$6.8 - (1+3+5+7+9) = 48 - 25 = 23$

11. A

$$\text{shaded area} = \left(e\sqrt{2}\right)^2 - \frac{\pi\left(e\sqrt{2}\right)^2}{4}$$

$$= 2e^2 - \frac{\pi \cdot 2e^2}{4} = e^2\left(2 - \frac{\pi}{2}\right) = e^2\left(\frac{4-\pi}{2}\right)$$

12. C

$$\frac{360°}{8} \cdot 3 = 135°$$

13. C

Square has four lines of symmetry.

14. E

Since the set contains on odd member of consecutive integers and the smallest integer is odd, any odd number in the set can be placed in the center.

15. A

f(x) is defined for all real numbers

16. D

$-3x + 2y = 4$
$\underline{-2/\;\; 5x + y = 2}$
$-13x = 0 \;\rightarrow\; x = 0$ and $y = 2$

17. C

In order to get the graph of f(-x), reflecting f(x) in the y axis will suffice.

18. D

$3\alpha = 90°$
$\alpha = 30°$
$80 = x + 30°$
$x = 50°$

19. E

sum of the digits must be a multiple of 3.
$27 + R = 2k \;\rightarrow\;$ R can be 0, 3, 6 or 9

20. D

reflection is $y = -(x^3 - 5x) = -x^3 + 5x$

21. C

$149.95 = 12.3n + 4.56 \;\rightarrow\; n = 11.82$
$\rightarrow$ 11 items can be produced

22. A

Given is a right triangle.

$$(2R)^2 = 5^2 + 12^2 \;\rightarrow\; R = \frac{13}{2}$$

area of the shaded region is

$$\pi\left(\frac{13}{2}\right)^2 \cdot \frac{1}{2} - \frac{5.12}{2} = 36.37$$

23. E

If $Sinx = Cosy$ then $x + y = 90°$ so y can be $\hat{1}$ or $\hat{3}$.

24. A

B(-1,5)

25. D

$AC^2 = 4^2 + 8^2 \;\rightarrow\; AC = 4\sqrt{5}$

$$f(c) = \frac{1}{\sin C} = \frac{1}{\dfrac{8}{4\sqrt{5}}} = \frac{\sqrt{5}}{2}$$

26. B

$$x^2 - 9 \neq 0 \Rightarrow x^2 \neq 9$$
$$x \neq \mp 3$$

27. A

$g(f(1.5)) = g(1.5^2-1) = g(2.25-1) = g(1.25)$

$= \sqrt{1.25} + 4 = 5.12$

28. D

if A is zero then y = -x

so B = xy = x.-x = $-x°$; B is either zero or negative

29. B

m = 1 + (9-7) = 3

n = 1 + (6-1) = 6

30. E

area ratio = $\dfrac{4}{5}$

➔ side ratio = $\dfrac{2}{\sqrt{5}}$ ➔ volume ratio =

$\left(\dfrac{2}{\sqrt{5}}\right)^3 = \dfrac{8}{5\sqrt{5}} = \dfrac{8\sqrt{5}}{25}$

31. C

Since $Sin^2x + Cos^2x=1$ and x is an acute angle,

$Cosx = \sqrt{1 - Sin^2x} = \sqrt{1 - c^2}$

32. D

$x^2-x-12 \geq 0$

$(x-4)(x+3) \geq 0$

33. D

P(-2) = 0 ➔ $-8 - 8 - 6b - 14 = 0$

-6b = 30

b = -5

34. C

f(3x-4) = 4x – 10

f(E) = 2

E = 3x -4 and 4x – 10 = 2

X = 3 ➔ E = 3.3-4 = 9 – 4 = 5

35. C

$(0.8)^2 = 0.64$

Area decreases by $\dfrac{1 - 0.64}{1} x100 = 35\%$

36. C

$x^2 = 10^2 + 24^2$ ➔ x = 26

$\dfrac{y}{26} = \tan 37° \Rightarrow y = 26.\tan 37° = 19.59$

$\Rightarrow Area = \dfrac{26 x 19.59}{2} = 254.7$

37. A

There are totally 9.9=81 possibilities but once she has tried out 80 attempts and even if she is unsuccessful the she will be sure that in the next attempts, she will find out her password.

38. B

$\begin{pmatrix} 12 \\ 2 \end{pmatrix} = \dfrac{12.11}{2} = 66$

39. C

2x + 100 = 180°

x= 40° ➔ Number of sides = $\dfrac{360°}{40°} = 9$

40. E

$-p = \dfrac{2-p}{p-1-3} \Rightarrow -p^2 + 4p = 2-p$

$p^2 - sp + 2 = 0 \Rightarrow p = 0.44 \text{ or } 4.56$

41. B

A,C,D are parallel to the given line so they will lead to no solution E is the same as the given line so it will lead to infinitely many solutions so answer is B.

42. D

He bought each of 100$; regular price for each is 175$.

He paid 16x100=1600$

He got 8 x 175 + 4 x 175 x 0.10 + 4 x 175 x 0.40 = 2170

So his rate of profit is (2170-1600)/1600 x 100 = 35.6 %

43. B

AB = BC = CA = $x\sqrt{2}$ so perimeter of ABC is

$3.x\sqrt{2} = 3x\sqrt{2}$

44. B

$-7< x < 15$ so -11 < x-4< 11 ➔ |x-4| < 11

45. B

I and III are not necessarily correct. II is correct all the time.

46. D

Mrs. White does not wear the White and Grey hat; therefore she has to wear the black hat.

47. C

5 – 12 – 13 is a right triangle and the hypotenuse is the side with length 13. The hypotenuse will be the diameter of the circle in question. Thus, 2R = 13 and R = 6.5

48. B

$\dfrac{12 x 18 x 24}{2 x 2 x 2} = 6x9x12 = 648$ Since each ball should be placed in a cube that is 2 inches on a side.

49. B

$\log \dfrac{\sqrt{a}}{b} = \log \sqrt{a} - \log b = \log a^{1/2} - \log b$

$= \dfrac{1}{2}\log a - \log b = \dfrac{1}{2}x - y$

50. A

$3^2+(2-1)^2=9+1=10<16$ ➔ Answer is A.

Model Test 2 – Solutions

1. A

$7z+28 = z-3$

$6z=-31$ ➔ $z=\dfrac{-31}{6}$ ➔ -5.17

2. E

$x=5$ ➔ $y = 2(5-3)^5 = 2 \times 2^5 = 64$

3. D

$\left[\dfrac{13}{3}\right] = 4 \quad \left[\dfrac{22}{3}\right] = 7 \quad \left[\dfrac{17}{3}\right] = 5$

The number of cubical boxes is 4 x 7 x 5 =140

4. C

$\dfrac{5.AB}{2} = 20$

$AB = 8$ ➔ $a=2-8 = -6$

5. B

$\left. \begin{array}{l} 2y = 3x + c \\ 2.\dfrac{-1}{2} = 3.3 + c \Rightarrow c = -10 \end{array} \right\} \begin{array}{l} 2y = 3x - 10 \\ 3x = 2y + 10 \end{array}$

6. D

$2.3 + 3.8 + 4.15 + 5.24 = 6 + 24 + 60 + 120 = 210$

7. E

$x^2 - y^2 = (x+y)(x-y)=11.7=77$

8. B

$2. (4 + 6 + 2) = 22$ cubes have point on only one face.

➔ $\dfrac{22}{60}.100 = 37\%$

9. D

$x^2 = 5^3$

$x= 5^{\frac{3}{2}}$

$x^{\frac{3}{2}} = 5^{\frac{3}{2}.\frac{3}{2}} = 5^{\frac{9}{4}} = 37.38$

10. D

$2\pi.\dfrac{7}{2}.12 = 263.89$

11. C

$\dfrac{A}{x} \quad \dfrac{B}{2x-10} \quad \dfrac{C}{2x}$

$(x-10).3=2x-10$ ➔ $3x-30 = 2x-10$

$x = 20$ ➔ $B= 2.20 -10 = 30$

12. D

Perimeter = $10 + 10 + \sqrt{7} + \sqrt{3}$

$+1 + \sqrt{7} - (\sqrt{3} -1)$

$= 22 + 2\sqrt{7} = 27.29$

13. B

$30^{\frac{2}{n}} = (2.3.5)^8 = 30^8$

$\dfrac{2}{n} = 9 \Rightarrow n = \dfrac{2}{8} = 0.25$

14. B

$P(3)=8 = 27 + 9t + 15 + 2$

$9t=-36$

$t=-4$

15. C

$2 \lozenge a = 2^a - a^2 < 0$

a cannot be -1 since it must be positive so a can only be 3.

16. C

$x - \dfrac{x}{4} - \dfrac{x}{5} = 110$

$\dfrac{11x}{20} = 110 \Rightarrow x = 200$

$200.\dfrac{1}{4} = 50$

17. B

$| x-x-x | = | -x | = -x$

18. E

$\theta = 130° - 60° = 70°$

19. C

length of arc ADC = $\dfrac{2\pi.5}{3} = \dfrac{10\pi}{3} \approx 10.5$

20. C

$\sqrt[5]{x^2} = 4$

$x^{\frac{2}{5}} = 4 \Rightarrow x = 4^{\frac{5}{2}} = 2^{2.\frac{5}{2}} = 2^5 = 32$

21. B

$\dfrac{2}{\sqrt[3]{7+1}} = \dfrac{2}{\sqrt[3]{8}} = \dfrac{2}{2} = 1$

22. C

$-4 \leq 3-x \leq 4$

$-7 \leq -x \leq 1$

$7 \geq x \geq -1$ ➔ $-1 \leq x \leq 7$

23. B

$$\frac{1}{2}\cdot\frac{1}{2}\cdot\frac{1}{2}\cdot\frac{1}{2}\binom{4}{3}=\frac{4}{16}=\frac{1}{4}$$

24. B

$$i^{14}+1=i^2+1=-1+1=0$$

25. D

A: 2^{-121} B: 2^{-122} C: 2^{-120}

D: 2^{-128} E: 2^{-125}

➔ 2^{-128} is the least

26. A

$x^2 + y^2 = R^2$ since the circle is centered at the origin

$2^2 + 5^2 = R^2$ ➔ $R^2 = 29$ ➔ $x^2 + y^2 = 29$

27. B

$5\alpha = 90°$

$\alpha = 18°$

$$\tan\alpha = \frac{4}{x}=\tan 18°$$

➔ $x = \dfrac{4}{\tan 18°}=12.31$

28. C

I. $A^{2+3}+1=$ odd + odd + odd = odd

II. $2(a+b)+1 =$ even + odd = odd

III. $a+2b+1 =$ odd + even + odd = even

29. C

$(Sinx + Cosx)° + (Sinx - Cosx)^2$

$= Sin°x + 2Sin^2Cosx + Cos°x$

$+ Sin^2x - 2SinxCosx + Cos^2x = 2(Sin^2x + Cos^2x)=2$

30. C

The lengths are all positive quantities, therefore either one pencil or two pencils are longer than 5 inches.

31. E

$$y = 2.\frac{x+1}{2}-3=x-2$$

y= x-2

y=0   ➔ x=2

32. B

$$A =\binom{11}{2}=\frac{11.10}{2}=55$$

B= 11-1 = 10

A-B = 55-10 = 45

33. C

$6a^2 = 2a^3$ so a = 3 and diagonal = $a\sqrt{3} = 3\sqrt{3}$

34. E

All statements are correct. Think of a pentagonal prism for instance. H has is Edges, 10 vertices and 7 faces.

35. D

$$\frac{a}{b} = \frac{3}{3+2} = \frac{3}{5}$$

$$\frac{a}{a+b} = \frac{3}{3+5} = \frac{3}{8}$$

36. C

(-6, 2) → (6 , 2) → (6, -2)

37. E

$$10.12x\left(1+\frac{115}{100}\right)^5 = 464.9$$

38. C

$$Sin^2\theta = 1- Cos^2\theta = 1-\frac{a}{b}$$

$$Sin\theta = \sqrt{1-\frac{a}{b}}$$

39. B

B is the correct position.

40. B

$x + 2 \geq 0$ and $x-1 \neq 0$

$x \geq -2$ and $x \neq 1$

which is equivalent to B.

41. D

$a^3 = 64$ ➔ a=4

$\pi.2^2.4=16\pi=50.27$

42. E

$\dfrac{160}{8} = 20$ so perimeter = $\left(20+8\sqrt{2}\right)2 = 62.63$

43. D

$(2-2)^2 + 2 \leq y \leq (4-2)^2 + 2$

$2 \leq y \leq 6$

44. D

$a_n = -2 + (n-1).6 = 6n - 8$ Which is equivalent to A.

45. C

$x = \sqrt{64 + 18} = \sqrt{82}$

Perimeter = $\sqrt{82}.2 + 6\sqrt{2} = 26.6$

46. C

If the equation is quadratic then $\Delta=0 = 25-4.(2p-1).4 = 0$ ➔ $p =\dfrac{41}{32}$

If the equation is not quadratic then 2p-1 = 0 ➔ $p =\dfrac{1}{2}$

In both cases we have one and only one real solutions the solutions set.

47. C

$\hat{AOB} = 120°$

➜ Shaded area = $\dfrac{\pi.10^2}{3} - \dfrac{10\sqrt{3}.5}{2} = 61.42$

48. C

$(a-2)^2 + y^2 = 9$ must have a single solution

So y must be zero ➜ $(a-2)^2 = 9$ ➜ $a-2 = \mp 3$

A= 2 ∓ 3 ➜ -1 or 5.

49. B

p ➜ q is equivalent to p'vq and q' ➜ p'.

➜ I and IV are correct.

50. C

$x^2 = 4.\ 16$ so $x = 2.4 = 8$ and $y = 20$; thus $x + y = 28$

Model Test 3 – Solutions

1. D

a- b =11

$4a - 4b = 4.11 = 44$

2. B

$f\left(\dfrac{2}{5}\right) = f(0.4)$

$= 5(0.4)^2 - 3(0.4)+4 = 3.6$

3. B

$210.3 = 630$ min = 10 hours & 30 min.

10 hours & 30 min after 8:00 AM is 6:30PM

4. C

$3^{n+3} = 243 = 3^5$

n+3=5 ➜ n=2

5. A

The distance from the center of the circle to the line is

R and $R = \dfrac{|4(1)-3(-2)+10|}{\sqrt{16+9}} = \dfrac{20}{5} = 4$

6. D

$2x^2 - 5x - 3 = (2x+1)(x-3)$

7. A

$x^2 = 9$; $x = -3$ since $x< 0$, thus $x^3 = (-3)^5 = -243$

8. B

$f(g(-1)) = f(-1) = |1-1| = 0$

9. D

$\dfrac{3}{5}.x^2 = \dfrac{3}{5}.16 = 9.6$

10. E

$2x^3+3x^2-11x-6 = 2x^3-11x^2+17x-6$

$14x^2-28x=0$

$14x(x-2)=0$ ➜ x=0 OR x=2

11. C

$x^2 - 2xy + y^2 = 25$

$x^2 + 2xy + y^2 = 49$

$\overline{}$

$-4xy = -24$ ➜ xy=6

12. B

from $\triangle ABE$ $m\left(\hat{ABF}\right) = 90-\theta$

from $\triangle FED$ $m\left(\hat{EFD}\right) = 90-\theta = m\left(\hat{AFB}\right)$

➜ $\triangle ABF$ is equilateral and AB = FA = 12

13. D

$x(y+1)$= odd ➜ x=odd and y=even

so I and II are correct.

14. E

$\left.\begin{array}{l} \dfrac{x}{3} = \text{Cos}\theta \\ y = \text{Sin}\theta \end{array}\right\}$ $\begin{array}{l} \text{Cos}^2\theta + \text{Sin}^2\theta = 1 \\ \left(\dfrac{x}{3}\right)^2 + y^2 = 1 \end{array}$

15. D

$\sqrt[4]{x^3} = 24 \Rightarrow x^{\frac{3}{4}} = 24 \Rightarrow x = 24^{(4/3)} = 69.23$

16. C

3y=2x-11

$y = \dfrac{2}{3}x - \dfrac{11}{3} \Rightarrow$ slope $= \dfrac{2}{3}$ the given line is

perpendicular to 3x+2y=8 since $y = \dfrac{-3x}{2} + \dfrac{8}{2}$ gives

the slope of $\dfrac{-3}{2}$.

17. A

Measure of angle EDC = (60° – 40°) / 2 = 10°

18. C

x-1=0 ➜ x=1

$2.1^5 + 4.1^3 - 1 + 1 = 2 + 4 = 6$

19. A

Assume that "Less than 6" means 5. Then the average would be

$\dfrac{42.20 + 26.18 + 22.15 + 14.8 + 8.5}{42 + 26 + 22 + 1448} = 16.01$. Therefore average cannot exceed 16.01.

20. B

$2x^2 + x-1 \leq 0$

$(2x-1)(x+1) \leq 0$

$-1 \leq x \leq \dfrac{1}{2}$

21. E
$2x + 3x + 4x + 5x + 6x = (5-2).180$
$20x = 540$
$x = 27 \quad \to 6x = 6.27 = 162°$

22. B
$n^2 - 2n = 3$
$n^2 - 2n - 3 = 0$
$(n-3)(n+1) = 0 \to n=3 \text{ OR } n= -1$

23. A
x^3 is negative
so x is negative

24. B
$\sqrt{2}Sin^2x + \sqrt{2}Cos^2x = 2Sinx$
$\sqrt{2}(Sin^2x + Cos^2x)2Sinx$
$\sqrt{2} = 2Sinx \Rightarrow Sinx = \dfrac{\sqrt{2}}{2}$
$x=45 \to tanx=1$

25. A
$4x+x^2=77$
$x^2+4x-77=0$
$(x+1)(x-7)=0$
$x=7$

26. C
$3.9(3)-2=16$
$3g(3)=18$
$g(3)=6 \to$ C is the correct answer since $3^2-3=6$

27. D
$x^2-y^2=45$
$(x-y)(x+y)=45$
$x-y=5$
$x+y=9$
$x-y=5$
$\to x=7$
$y=2 \to \dfrac{2x}{y} = \dfrac{14}{2} = 7$

28. A
$2x+3x+4x=360°$
$9x=360^3$
$x=40°$ greatest ext angle = 160° $\to$ smallest int angle = 20°

29. D
$a+b=-2+3=1$

30. D
$\dfrac{7.72+6.69}{13} = 70.62$

31. D
$-2 \le \dfrac{3x-1}{5} \le 1$ so $-10 \le 3x-1 \le 5$
$-9 \le 3x \le 6$ and $-3 \le x \le 2$

32. D
$y = \sqrt{x^2} +1 = |x| +1 ==> D$ is correct

33. E
(a,b) can be (-1, 4), (-3,-4) or (11,4)
so a+b can be 3, -7 or 15

34. B
$f(s)=-25 \to f(-5)=-25 \to$ (-5, -25) is on the same graph.

35. C
$y=-2(x^2-2x)-1$
$= -2(x-1)^2+1$
$\to$ max value of f(x) is 1.

36. E
$f(101.5) = 3$
$f(34.4) = 6 \qquad \to 346-0=9$
$f(133.6)=0$

37. E
$\left(\dfrac{f}{g}\right)(x) = \dfrac{x^2 + 4x - 5}{-x^2 - x + 2} \Rightarrow$
$-x^2-x+2=0$
$x^2+x-2=0$
$(x+2)(x-1)=0$
$x=-2$ & $x=1$ are not in the domain of x.

38. B
$\dfrac{n}{75} = tan27° \Rightarrow h = 75 tan27° = 38.21$

39. D
$V = \dfrac{1}{2}.(\pi.5^2.12)$
$= 150\pi$
$=471.24$ (Since we rotate for 180°, we get half of a cylinder).

40. A
$\dfrac{a^2\sqrt{3}}{4} = b^2 \Rightarrow \dfrac{a}{b} = \sqrt{\dfrac{4}{\sqrt{3}}} = 1.52 \Rightarrow \dfrac{3a}{4a} = 1.14$

41. B
The possible remainders are 1,2,3, ..., 12 and the ones that are prime are 2,3,5,7,11
$\to$ Required probability is $\dfrac{5}{12}$

42. C
$\left.\begin{array}{l} a = k.v^2 \\ 6 = k.46^2 \\ 12 = k.v^2 \end{array}\right\} \quad \dfrac{6}{12} = \left(\dfrac{46}{v}\right)^2 \Rightarrow v = 65m/s$

43. E

$a\sqrt{3} = d$

$a = \dfrac{d}{\sqrt{3}}$

$6a^2 = 6 \cdot \left(\dfrac{d}{\sqrt{3}}\right)^2 = 6 \cdot \dfrac{d^2}{3} = 2d^2$

44. B

$3\left(\log_7 4\right)^2 = 1.52$

45. B

$Cos\alpha\left(\dfrac{Sin\alpha}{Cos\alpha} + \dfrac{1}{\dfrac{Sin\alpha}{Cos\alpha}}\right) = Sin\alpha + \dfrac{Cos^2\alpha}{Sin\alpha}$

$= \dfrac{Sin^2\alpha + Cos^2\alpha}{Sin\alpha} = \dfrac{1}{Sin\alpha}$

46. A

$\pi R^2 = 16\pi$

$R = 4$

Shaded area $= \dfrac{1}{4}\left(8^2 - 16\pi\right)$

$= \dfrac{1}{4}\left(64 - 16\pi\right) = 16 - 4\pi$

47. A

Those who checked exactly 3 books = These who checked of least 2 books – Those who checked of least 3 books = 16 – 14=2

48. B

Total number of people = Those who checked none or at least 1 = 4 + 20 = 24

49. B

By Euclid's theorem

$DC^2 = AD \cdot DB = 49$

$DC = 7$

50. D

Triangles ABD and BCD have the some base and height; triangles BCD and CDE also have the some base and height. So area of $\overset{\triangle}{ABD}$, area of $\overset{\triangle}{BCD}$ and area of $\overset{\triangle}{CDE}$ are all equal.

Model Test 4 – Solutions

1. E

9x=12x-12➔3x=12 so x=4 ➔ 2x=8

2. A

$\dfrac{\frac{x}{9} = x \cdot \frac{x}{9} = \frac{x^2}{9}}{x}$

3. E

(1 - -1)(-1-1)=2(-2)=-4

4. A

n-1=341

n=5   m+1=6

                                      m=6

n-m = 5-5=0

5. B

(a+2+b)(a+2-b)=(a+2)²-b²

6. E

y=0 ➔ $\dfrac{2}{3}x = \dfrac{4}{5}$ and $x = \dfrac{12}{10} = \dfrac{6}{5}$

7. B

3xy=5 so 9x²y²=(3xy)²=5²=25

8. D

α+β=180°

9. C

x²-y²= 33 = (x-y)(x+y)

$33 = 3(x + y$ $x + y = 11$⎫ $2x = 14$

$x + y = 11$ $x - y = 3$⎬$x = 7 \Rightarrow y = 4$

10. B

$\sqrt{\sqrt[3]{x}} = \dfrac{1}{2} \Rightarrow x = \left(\dfrac{1}{2}\right)^6 = \dfrac{1}{64}$

11. C

The digits that must be used are 3, 2, 1 and 0. And there are 3·3·2·1= 18 such numbers.

12. A

$\dfrac{A}{3x} \dfrac{B}{x} \dfrac{C}{\frac{3x}{2}}$ 3x=216 and x=72 so $|B-C| = \dfrac{72}{2} = 36$

13. D

The triangle before the rotation is equilateral and acute.

14. C

6x=y

2x-y = 2x − 6x = -4x

15. E

$f\left(\dfrac{1}{2}\right) = \dfrac{1}{\left(\frac{1}{2}\right)^2} = 4$

16. A

$20.10^n = 32.625$

$10^n = 1000 \Rightarrow n = 3$

17. E

$|4-1| \geq 3$ is correct

$3 \geq 3$

18. B

Required probability is $\dfrac{\frac{4.4}{2}}{4.5} = \dfrac{8}{20} = \dfrac{2}{5} = 40\%$

19. C

Perimeter of the shaded region is

$\dfrac{2.\pi.6}{3} + \dfrac{6}{2}.\sqrt{3}.2 = 4\pi + 6\sqrt{3} = 22.96$

20. B

$\tan\theta = A = \dfrac{A}{1}$

$\sin\theta = \dfrac{A}{\sqrt{1+A^2}}$

21. D

Number of possible ID codes are 5.4.8=160

22. A

$\dfrac{1}{2}.\dfrac{1}{2}.\dfrac{1}{2}.4$ since HHT, HTH, THH and HHH $= \dfrac{1}{2}$

23. D

Perimeter of the shaded region is 2.6=12

24. B

$i + i^2 + i^3 + i^4 = i - 1 - i + 1 = 0 \Rightarrow$ min value of k is 4.

25. C

$2.\pi.2.5.\dfrac{360° - 63°}{360°} = 12.96 \approx 13$ is the best answer.

26. C

$m^2 = A$

$n^2 = B$

A-B=$|m^2 - n^2|$=m+n $\rightarrow$ $\left|\dfrac{m^2 - n^2}{m+n}\right| = 1 \Rightarrow |m-n| = 1$

$\rightarrow$ answer is C.

27. E

$40 + 5x < 7x$

$40 < 2x$

$20 < x$

28. C

$x^3 = y^3 \rightarrow$ x=y

$|x| = |y| \rightarrow x = \pm y$

$\sqrt{x} = \sqrt{y} \rightarrow$ x= y

29. C

d in feet is $\dfrac{1}{6}.(2^2 - 4.2 + 7).3 = \dfrac{9}{6} = \dfrac{3}{2}$ ft.

30. B

$|OP| = \sqrt{6^2 + 2^2} = \sqrt{40}$

$|PS| = \sqrt{2^2 + 6^2} = \sqrt{40}$

$|OP| = |PS|$

31. E

x intercept of n is always less than that of m and it can be negative, zero or positive.

32. D

$(2\cos^2\theta + 2\sin^2\theta - 5)^4 = (2-5)^4 = (-3)^4 = 81$

33. E

A plane extends flat in all directions so plane ACG also contains point E.

34. C

$\dfrac{\pi.x^2}{\frac{x.x}{2}.2} = \dfrac{\pi x^2}{x^2} = \pi$

35. C

$\dfrac{6}{18} = \dfrac{3}{x+4} = \dfrac{x}{3+y}$

$\rightarrow$ x=s and   y=12 $\rightarrow$ $\dfrac{x}{y} = \dfrac{5}{12}$

36. B

$(x-1)^2+(y+2)^2=25$ and y=x

$\rightarrow$ $(x-1)^2+(x+2)^2=25$

$2x^2+2x-20=0$

$x^2+x-10=0$

x=2.7 or -3.7

37. B

$1.205 \times \left(1 - \dfrac{21}{100}\right)^5 = 370.79$

38. A

The y coordinates will be -4, -4 and, -7 and their sum is -15.

39. A

$\overset{\Delta}{ABC}$ is a right triangle if and only if the center of the circle lies on AB, BC or AC.

40. B

$AC = \sqrt{1^2 + 4^2 + 2^2} = \sqrt{1 + 16 + 4} = \sqrt{21}$

41. E

$a+b+c+d+90° = (5-2).180 = 540°$

$a+b+c+d = 450°$

$450-250 < b+d < 450-170$

$200 < b+d < 280$

42. C

$x^3-1 = 3$

$x^3 = 4$

$x = 1.59$

43. D

The graph of $f(x)$ posses the horizontal line test therefore $f(x)$ is invertible. $fof^{-1} = I$ where I is the identity function therefore $I(2) = 2$

44. D

$DE = EC = \sqrt{4^2 + 4^2 + 2^2} = \sqrt{36} = 6$

Perimeter of $D\overset{\Delta}{E}C$ $646+4 = 16$

45. B

$a_n = 15 + (n-1).4 = 4n + 11$

46. D

$a = 2b+1$ and $b = 2c+1$

$a = 2(2c+1)+1 = 4c+3$

47. B

The areas are the some since the circle is shifted to the right and the dimensions of the shaded region do not change.

48. B

$AB = \sqrt{(2.5\pi)^2 + 12^2} = 14.34$ in $= 1.20$ ft

49. B

$196 \begin{vmatrix} 81 & 27 & 9 & 3 & 1 \\ 2 & 1 & 0 & 2 & 1 \end{vmatrix}$ EDCBA = 21,021 and the

remainder when EBCBA is divided by 4 is 1.

50. D

$7x = \sqrt{(3x)^2 + (2x)^2 + 4^2}$

$49x^2 = 9x^2 + 4x^2 + 4^2$

$h^2 = 36x^2$

$h = 6x$

Model Test 5 – Solutions

1. C

$\sqrt{196-169} = 3^n \Rightarrow \sqrt{27} = 3^n$

$\sqrt[2]{3^3} = 3^n$

$3^{3/2} = 3^n \Rightarrow n = \dfrac{3}{2} = 1.5$

2. E

$\sqrt[2]{\dfrac{\left(5^{-1}\right)^{-1} : \left(5^1\right)^2}{5^{-3}}} = \sqrt[2]{\dfrac{5^1.5^2}{5^{-3}}} = \sqrt[2]{5^6} = 5^{6/2} = 5^3 = 125$

3. E

Since the integers are equally spaced, median and mean are equal. Since all numbers are positive, mean is positive. Each number occurs once so there is no made.

4. D

$64.8 \times 10^{-6} = 0.0000648$

There are 4 zeros between the decimal point and the first nonzero digit.

5. D

$ab > 0$ and $ab < 1$ ➔ best answer is D

6. E

E is false because $(a+b)^4 \neq a^4 + b^4$

7. D

$5.4-2.-2 = 24$ and 24 is not less than 21 ➔ answer is D.

8. B

$100.\dfrac{20}{100}.\left(1+\dfrac{12}{100}\right)^3 + 100.\dfrac{80}{100} = \$108.1$

9. D

$f(A) = \dfrac{A^{14}}{A^8} = A^{14-8} = A^6 \Rightarrow f(3.3) = 3.3^6 = 1291.5$

10. E

- $4a$ is even so b is odd. Therefore $2a+b$ is definitely odd.

11. A

$\left.\begin{array}{l} E_1 = mc^2 \\ E_2 = m.(2c)^2 \end{array}\right\} \dfrac{E_1}{E_2} = \dfrac{1}{4} \Rightarrow E_2 = 4E_1$

12. A

$\tan A\overset{\Delta}{C}B = \dfrac{6}{x} = 2 \Rightarrow x = 3$

13. A

$\left.\begin{array}{l} \sqrt{x-\dfrac{2}{3}} > \sqrt{x-\dfrac{3}{2}} \\ \sqrt{x-\dfrac{3}{4}} > \sqrt{x-\dfrac{4}{3}} \end{array}\right\}$ LHS > RHS ➔ no solution for x.

14. A

Correct answer is given in A.

15. C

$(x-4)^2 + (y-3)^2 < 2^2$ represent the required region.

16. B

$\left.\begin{array}{l}F(a,b)\\G(c,d)\end{array}\right\}$ shaded area = $c^2-ab-(c-a).d$

17. D

$-3\sin x = 1-2\sin^2 x-2\cos^2 x$

$-3\sin x = 1-2(\sin^2 x+\cos^2 x)$

$-3\sin x = 1-2=-1 \Rightarrow \sin x=\dfrac{1}{3}$

18. D

The difference between two consecutive unitary numbers is always 9.

19. E

$d_1 = r_2 \Rightarrow 2r_1 = r_2$

$\dfrac{\pi r_1^2}{\pi r_2^2} = \dfrac{\pi r_1^2}{\pi.4r_1^2} = \dfrac{1}{4}$

20. E

Domain is the set of all possible x values so answer is E.

21. D

$2 < x^2 < 14$

$\sqrt{2} < |x| < \sqrt{14} \Rightarrow \sqrt{2} < x < \sqrt{14}$ or $-\sqrt{14} < x < -\sqrt{2}$

Therefore x cannot be $-\sqrt{5}-\sqrt{3}$

22. E

$5x + 13x =90$

$18x=90 \Rightarrow x=5° \Rightarrow 13x = 65$

23. D

$P\overset{\triangle}{C}D$ is a right triangle so PL= $\sqrt{10^2 - 6^2} = 8$

24. A

ABCD is a rectangle and AC is the diameter.

25. B

$v_2 = \dfrac{1}{3}\pi(3r)^2.h = \dfrac{1}{3}\pi.9r^2h = 9\left(\dfrac{1}{3}\pi r^2h\right) = 9v_1$

26. C

$t^2=-x \Rightarrow y=2(-x)+1=-2x+1$

$y=-2x+1$ and $x \le 0. \Rightarrow$ this is a portion of a line.

27. C

Since $\theta = \beta+\sigma$, $\theta > \beta$

28. D

Slope of BC is negative since BC represents a decreasing line.

29. A

$|3x-6| < -2$

and $|3x-6|$ can never be negative. So the solution set is empty.

30. C

$4x^2 - 6x + 2c = 2ax^2 - 2bx - 4ax + 4b$

$4x^2 - 6x + 2c = 2ax^2 + x(-2b-4a) + 4b$

$2a= 4 \quad -6 = -2b - 4a$

$a=2 \quad -6 = -2b - 8 \quad \Rightarrow -2b = 2 \quad b=-1$

$2c=4b = -4$

$c = -2 \quad \Rightarrow 2—1—2 = 2+1+2=5$

31. E

$55^{22} = 1.94 \times 10^{38} \Rightarrow$ it is 39 digits.

32. E

Reflecting a relation across the origin means rotating it 180° about the origin. So correct answer is E.

33. C

The students that study Spanish may include the ones who study French and German. So at most 2200-800 = 1400 students study none of these languages.

34. E

$2y = x- 10 \qquad y = -ax + 5 \qquad \dfrac{1}{2}.-a = -1$

$y = \dfrac{1}{2}x-5 \qquad \Rightarrow a=2$

35. A

$P(P|F)=\dfrac{5}{21}$

36. E

Perimeter = 2(14+11)=50

37. C

In the 1st pattern, there are 2 unit line segments
In the 2nd pattern, there are 6 unit line segments
In the 3rd pattern, there are 12 unit line segments
In the 4th pattern, there are 12+8=20 unit line segments

38. D

$\dfrac{a^2 - 2bc - 2ac - b^2}{a+b.1} = \dfrac{a^2 - b^2 - 2c(a+b)}{a+b} =$

$\dfrac{(a-b)(a+b) - 2c(a+b)}{a+b} = \dfrac{(a+b)(a-b-2c)}{a+b} = a-b-2c$

39. E

$x=80-10-30=40°$

40. D

$2^7.2^{12} = 2^{19}$

41. D

$a^4 < b^4, a^2 < b^2$ and $|a| < |b|$ are all equivalent. But $a^4 < b^4$ cannot imply $a < b$.

42. C

When x = 0 $y_3 = 8$, $y_4 = \dfrac{1}{2}$ and $y_5 = 9$

➔ Three of the functions have positive y intercept.

43. A

The midpoint of (1, 7) and (a, b) is the center (-3, 4), thus (1 + a)/2 = -3 and (7 + b)/2 = 4, so (a, b) = (-7, 1).

44. E

C reads 75 pages in 3 hours and E reads 125 pages in 5 hours so each of them reads 25 pages in one hour.

45. D

$\sin(90° + \theta) = \cos\theta = \sqrt{1 - 0.38^2} = 0.92$

46. C

$R^2 = 5^2 + (12 - R)^2$ so $R^2 = 25 + 144 - 24R + R^2$

And 24R = 169 which gives $R = \dfrac{169}{24} = 7.04$

47. D

$$\left(\frac{\dfrac{1}{a}+\dfrac{1}{b}}{\dfrac{1}{a}-\dfrac{1}{b}}ab\right)^{-1} = \left(\frac{\dfrac{1+b}{ab}}{b-a}\right)^{-1} = \left(\frac{a+b}{b-a}\right)^{-1} = \frac{b-a}{b+a}$$

48. C

$AB = 4\sqrt{2} = 5.68$

49. B

$\left.\begin{array}{l} x = 0 \Rightarrow y = \mp3 \\ y = 0 \Rightarrow x = \mp2 \end{array}\right\}$ the relation represents as ellipse of

minor and major axis length equal to 4 and b respectively.

50. C

$5.4.4.4.4.4.4.4.4.3 = 15.4^8$

For the first region there are 5 possibilities. For the 2nd though the 9th regions there are 4 possibilities. For the last one there are 3 possibilities.

Model Test 6 – Solutions

1. C

$\dfrac{3\sqrt{10}}{2\sqrt{2}} = \dfrac{3}{2}\sqrt{5}$ is irrational

2. A

$2^{x-1} = 2^{-2}$ so x-1 = -2 and x= -1

3. D

midpoint is $\left(\dfrac{-3+3}{2}, \dfrac{4-4}{2}\right) = (0,0)$

4. E

$f(g(2)) - g(f(2)) = f((2-1)^3) - g(2^3-1)$
$= f(1) - g(7)$
$= 1-1 - (7-1)^3 = 0-6^3 = -216$

5. E

$\dfrac{1}{-1+\sqrt{3}} = \dfrac{1}{\sqrt{3}-1} \cdot \dfrac{\sqrt{3}+1}{\sqrt{3}+1} = \dfrac{\sqrt{3}+1}{3-1} = \dfrac{1+\sqrt{3}}{2}$

6. D

4 + 18 – x + 12 = 26
34 – x = 26
x=8

7. D

$x+y^3 = z^3y^3$
$x= z^3y^3-y^3=y^3(z^3-1)$

8. D

250, 252, 254,…, 400

Numbers of terms = $\dfrac{400-250}{2}+1 = 76$

9. A

$4y - 4x + 2z = -4(x-y)+2z$
$= -4.2 + 2.3$
$= -8 + 6 = -2$

10. C

$(-1)^5 - 1 - (1^5 - 1) = -1-1-(0) = -2$

11. C

$\left(x^2 - y^2 - 1\right)\left(\sqrt{x^2 + y^2} - 2\right) = 0 \Rightarrow x^2 + y^2 - 1 = 0$ OR

$\sqrt{x^2 + y^2} - 2 = 0$ so $x^2+y^2=1$ OR $x^2+y^2=4$

➔ Two circles centered of the origin with radii of 1 and 2.

12. E

$2.34(0.65\lceil 5.5 \rceil + 1) = 2.34(0.65 \times 6 + 1) = \$11.47$

13. A

$\tan A = \dfrac{\sin A}{\cos A} \Rightarrow \cos A = \dfrac{\sin A}{\tan A} = \dfrac{0.3090}{0.9511} = 0.325$

14. E

$5y = -\sqrt{5}x + 2\sqrt{5}$

$y = \dfrac{-\sqrt{5}}{5}x + \dfrac{2\sqrt{5}}{5} \Rightarrow m = \dfrac{-\sqrt{5}}{5} = -0.45$

15. D

$\log 4.12^a = \log 8.26^b$

$a\log 4.12 = b\log 8.26 \Rightarrow \left(\dfrac{a}{b}\right)^2 = \left(\dfrac{\log 8.26}{\log 4.12}\right)^2 = 2.22$

16. E

Median is the average of the two middle numbers and it is odd. Since terms are equally spaced median equals the mean. Since every term appears exactly once, there is no made.

17. E

Let p=5 and q=3 then 2p-q=10-3=7 and 7 is prime.

18. C

ABC is a right triangle. Therefore

$\frac{3x \cdot 4x}{2} = 6x^2$; the area is a multiple of 6.

19. B

Tibet will select 3 boys out of the remaining 8 boys and 4 girls (since Sevda is already invited) out of the 7 remaining boys. So answer is B.

20. D

For p → q, q' → p' is the contrapositive.

21. A

$\theta + 2x = 180°$

$2x = 180° - \theta$

$x = 90° - \frac{\theta}{2}$

$\text{Sin}x = \text{Sin}\left(90° - \frac{\theta}{2}\right) = \text{Cos}\frac{\theta}{2}$

22. D

$y = x^2 - 12$

$x = y^2 - 12$

$y^2 = x + 12$

$y = \mp\sqrt{x + 12}$

23. A

min AB = 9 and max AB = $\sqrt{9^2 + 12^2} = 15$

→ max − min = 15-9 = 6

24. D

$f\left(\frac{x-1}{2}\right) = 3x - 1$

$\frac{x-1}{2} = t \Rightarrow x = 2t + 1 \Rightarrow f\left(\frac{2t+1-1}{2}\right) = 3(2t+1) - 1$

f(t)=6t+2=2(3t+1)

25. B

p(A or B) = P(A) + p(B) − p(A and B)

$= \frac{1}{\underset{(8)}{3}} + \frac{1}{\underset{(6)}{4}} - \frac{1}{\underset{3}{8}} = \frac{11}{24}$

26. D

ac-bc=0

c(a-b)=0

c=0 or a-b=0 → c=0 or a=b

27. D

$\left.\begin{array}{l} a + 4d = 12 \\ a + 49d = 102 \end{array}\right\}$ 45d = 90

d = 2

a=12-8=4

28. C

f(x) must open downward and it must have no zeros. So a<0 and b²-4ac < 0

29. E

$\frac{3}{6} = \frac{2}{-A} \Rightarrow -3A = 12$

A= -4

30. B

$\frac{1}{3}(x + y) = 35$

x+y=105 → z=180-x-y = 180-105=75°

31. D

If x=1 then $4+1 > 3^1$

5 > 3

32. E

|x-1| < 4

-4 < x-1 < 4

-3 < x < 5

-3 < x < 5 does not imply |x| < 2 that means -2 < x < 2

33. C

Sum of an interior and exterior angle is 180° → the required sum is 7 . 180°.

34. C

Shift f(x) up. Answer is C.

35. D

$x^3 = 2^{15}$ → $x = 2^{15/3} = 2^5 = 32$

36. E

E satisfies all points, therefore answer is E.

37. C

Average Velocity $= \dfrac{48 + 48}{\dfrac{48}{24} + \dfrac{48}{16}} = \dfrac{96}{2+3} = \dfrac{96}{5} = 19.2$

38. A

Reflect f(x) in the x axis then in y axis. Answer is A.

39. D

Shaded area = Outer area − Inner area
= (a+2w)(b+2w)-ab

40. C

In group 3; 9,8,7,7 are the numbers. Median is 7.5

and mean is $\dfrac{9+8+7+7}{4} = 7.75$; So mean is greater then median.

41. B

Domain is $a < x < e$; g=m is also included in the domain.

42. B

$a^2 + 4ab + 4b^2 \geq a^2 - 4ab + 4b^2 + 18$

$8ab \geq 18$

$ab \geq 2.25$ and $a \leq b$. Thus for b to be least $a=b$ and $ab= 2.25$ ➔ $b=1.25$

43. B

Coordinates of C are needed to determine the altitude and point C is unknown.

44. D

$11000 - 4000 = 7000$

45. E

In 1993 $\dfrac{6000 + 7000}{2} = 6500$

In 1995 $\dfrac{7000 + 9000}{2} = 8000$

Percent increase = $\dfrac{8000 - 6500}{6500} \cdot 100 = 23\%$

46. A

3,4,5 is 2 right triangle

$3-r+4-r = 5$

$2r = 2$ ➔ $r=1$ ➔ $d=2$ ft = 24 in

47. C

Triangle ABC and EDC are similar but they are neither congruent nor isosceles.

48. B

$\dfrac{4}{\sqrt{16 + x^2}} + \dfrac{4}{x} = \dfrac{32}{15}$

$x= 3$ satisfies the equation.

49. D

III equals = $f\left(\dfrac{-1}{x}\right) = \dfrac{-1}{x} + x = x - \dfrac{1}{x}$; all others equal

$-x + \dfrac{1}{x}$.

Answer is D.

50. C

$-8 - 8 - 6p - 14 = 0$

$-6p = 30$

$p = -5$

Model Test 7 – Solutions

1. C

$\dfrac{1}{t + 3} = 6 \Rightarrow 6t + 18 = 1$ so $6t= -17$ and $t= -17/6$

$\dfrac{1}{t - 3} = \dfrac{1}{\dfrac{-17}{6} - 3} = \dfrac{6}{-35}$

2. B

$\dfrac{\text{decided}}{120 - x} \quad \dfrac{\text{undecided}}{x}$

$(120 - x) \cdot \dfrac{80}{100} + 18 + x = 120$

$96 - 0.8x + 18 + x = 120$

$0.2x = 6$

$x = \dfrac{6}{0.2} = \dfrac{60}{2} = 30$

3. E

$x \cdot \dfrac{1}{7} + 6 + x \cdot \dfrac{5}{7} = x$

$6 = \dfrac{x}{7} \Rightarrow x = 42$

4. E

$\dfrac{1}{5 - 2\sqrt{3}} \cdot \dfrac{5 + 2\sqrt{3}}{5 + 2\sqrt{3}} = \dfrac{5 + 2\sqrt{3}}{13}$

5. C

The product is odd so each of the integers must be odd. The product is negative so one, three or all of the integers must be negative.

6. D

$2-x=3x-6$ OR $2-x=-3x+6$

$-4x=-8$ $2x=8$

$x=2$ $x=4$

7. A

$x+3=k(y+3)$ $x+3=2(y+3)$

$9+3=k \cdot (3.3)$ $x+3=2.(6+3)$

$k= \dfrac{12}{6} = 2$ $x=18-3=15$

8. D

$10t - 9 = \dfrac{1}{3} \cdot AT$

$AT=30t-27$ ➔ $ST = 30t - 27 + 101 - 9 = 40t - 36$

9. C

(4,0) is not a solution because. It does not satisfy the inequality given that x=4 and y=0

10. D

$5t+60°+30°=360°$

$5+t=270°$

11. C

A must be nonzero, otherwise f(x) represents a line resulting in the fact that f(x) and −f(x) intersect at one point only. Discriminant must be negative so f(x) does not intersect the x axis resulting in the fact that f(x) and −f(x) do not intersect.

12. E
A two digit prime number is always odd. So 30 and $3a+2$ are odd and $3a+1$ is even. So $3a(3a+1)(3a+2)$ is even.

13. B
$\sqrt{2.2^2 + 2} = 2.615 \approx 2.62$

14. B
$DC = b+4-(b-1) = 5$
So coordinates of A are $(1, -1+5)=(1,4)$

15. C
One side of eq. triangle = x
One side of regular hex. = y
$$\left(\frac{x^3\sqrt{3}}{4}\right)\bigg/\left(\frac{6.y^2\sqrt{3}}{4}\right) = \frac{2}{3} \Rightarrow \frac{x^2}{6y^2} = \frac{2}{3} \Rightarrow \frac{x^2}{y^2} = 4$$
$$\frac{x}{y} = 2 \Rightarrow \frac{3x}{6y} = \frac{1}{2}.2 = 1$$

16. D
Similarity ratio $= \dfrac{6}{8} = \dfrac{a--4}{4}$ so $a+4 = \dfrac{24}{8} = 3$
$a=-1$
$$\frac{6}{8} = \frac{b--7}{10-7}$$
$$\frac{3}{4} = \frac{b+7}{3} \Rightarrow b+7 = \frac{9}{4} \Rightarrow b = \frac{9}{4} - 7$$
$$a+b = -1 + \frac{9}{4} - 7$$

17. C
$12f15 = \dfrac{60}{3} = 20$

18. E
$x^2 < 81$
$|x| < 9$
$-9 < x < 9$ ➔ x can be -8, -7,...,7, 8 ➔ sum = 0

19. C
$\dfrac{\frac{BD}{2}}{10} = \text{Sin}60°$
➔ $90 = 20\text{Sin}\,60° = 17.32$

20. B
$40 + x.\dfrac{20}{100} = x$
$40+x=5x$
$4x=40$ ➔ $x=10$

21. D
$\dfrac{-16}{2} = -8$ $\dfrac{-16}{8} = -2$
$\dfrac{24}{2} = 12$ $\dfrac{-16}{2} = -8$ $\Rightarrow -8 < \dfrac{m}{n} < 12$

22. E
AFCD and BCDE are not necessarily parallelograms.
➔ x cannot be determined.

23. D
$\dfrac{3+8+10+x+y}{5} = 14.2$
$x+y = 71-21 = 50$

24. E
Perimeter $\quad = 10 + 10 + 2.\pi.5$
$\qquad\qquad\qquad = 20 + 10\pi$
$\qquad\qquad\qquad =51.4$

25. C
$\left.\begin{array}{l} AC\,\text{Cos}x = 15 \\ AC\,\text{Sin}x + AC = 25 \end{array}\right\}$ $\dfrac{\text{Cos}x}{\text{Sin}x + 1} = \dfrac{3}{5}$
$25\text{Cos}^2x = 9\text{Sin}^2x + 18\text{Sin}x + 9$
$25(1-\text{Sin}^2x) = 9\text{Sin}^2x + 18\text{Sin}x + 9$
$34\,\text{Sin}^2x + 18\text{Sin}x - 16 = 0$
$17\text{Sin}^2x + 9\text{Sin}x - 8 = 0$
$17\text{Sin}x = -8$
$\text{Sin}x = 1$
$\text{Sin}x = \dfrac{+8}{17}$ Or Sinx= -1

26. C
Cost of 5 cars for 3 days= 5.f(3)

27. B
PR=10.1   QT=14.7   ➔ B may be false

28. E
$A(1,-2)$ is reflected over y axis to get B which is the intersection of m & r.

29. E
The C's that are A's are also B. So E is false.

30. C
$ab+c=a$
$c=a-ab$
$c=a(1-b)$➔ is a multiple of a.

31. C
CBEA11
1.26.26.14 = 9464

32. C
$6 < \sqrt{x-2} + 2 < 11$ so $4 < \sqrt{x-2} < 9$
$16 < x-2 < 81$
$18 < x < 83$

33. A
AB > OA ➔ CD > OA

34. B
$\dfrac{n(n-1)}{2} = 66 \Rightarrow n(n-1) = 132 = 12.11$
$n=12$

(218)

35. C

ABC is obtuse.

36. E

$i^{35} + i^{36} + i^{37} = i^3 + 1 + i^1 = -i + 1 + i = 1$

37. E

$8^2 + 11^2 > 12^2$

$64 + 121 > 144 ==$ greatest angle is acute so triangle is acute.

38. C

required ratio $= \dfrac{\pi.(\sqrt{2})^2}{\pi.1^2} = 2$

39. C

ratio of areas = (ratio of sides)^2 = (4:1)^2 = 16:1

40. B

$x^2 + 5x -2 = 2x + 8$

$x^2 + 3x - 10 = 0$

$(x+5)(x-2) = 0$

x=-5	x=2
y=-2	y=12

➔ (-5,-2) and (2,12) are the points of intersection

➔ $d = \sqrt{(-5-2)^2 + (-2-12)^2} = 15,65$

41. D

FHD is a right triangle whose sides are 4, $4\sqrt{2}$ and $4\sqrt{3}$.

42. C

$44000\left(1 - \dfrac{3}{100}\right)^9 = 33450$

43. E

Points above the parabola and below the line, including the points on the parabola but not those on the line are given in E.

44. E

$\dfrac{l}{5} = \dfrac{2}{\sqrt{3}} = \dfrac{x}{t}$ therefore $x = \dfrac{2t}{\sqrt{3}}$

45. E

$2 < \log_5 < 3$

$25 < x < 125$

46. C

99th term is -4 + 98 . 7 =682

47. E

In the n'th triangle the numbers are 1,2,3,....,n^2

➔ their sum is $\dfrac{n^2(n^2+1)}{2}$

48. D

at 20°C $\quad$ VI $= \dfrac{19}{9} = 2.11 \approx 2$

49. E

$\dfrac{6}{5} = 1.2 \qquad \dfrac{11}{6} = 1.83$

$\dfrac{13}{10} = 1.3 \qquad \dfrac{19}{9} = 2.11$

median of 1.2, 1.3, 1.35, 1.83 and 2.11 is 1.35 ➔ 25°C

50. E

$\dfrac{11,793}{1,972} = 5,98$ gr/m³ ≈ 6 gr/m³ ➔ 10°C

$\dfrac{11gr}{m^3} = \dfrac{x}{1972x2} = x = 11x1972x2 = 43384$gr $\approx$ 43400gr

Model Test 8 – Solutions

1. B

4. $\dfrac{1}{p^2} = 1 \Rightarrow p^2 = 4 \Rightarrow p = \mp 2$

2. D

$\dfrac{2x}{3} = \dfrac{3x}{2} \Rightarrow 4x = 9x \Rightarrow 5x = 0$

x=0

3. B

$\sqrt[5]{x} = 2.07$

$x = 2.07^5 = 38 \Rightarrow \dfrac{160}{x} = 4.21$

4. C

(-3, -7) when reflected across the x axis becomes (-3, 7)

5. D

41000=4.1x10^4

➔ k=4

6. E

$\left(\sqrt{7} - \sqrt{3}\right)\left(\sqrt{7} + \sqrt{3}\right) = 7 - 3 = 4$

7. D

$a^2 = 16$

$3b = 6 \qquad 2a^2 - 5 + 3b - 4c = 2.16 - 5 + 6 - 4.\dfrac{4}{3} = 27.7$

$c = \dfrac{4}{3}$

8. C

17b = 2a $\quad$ 5a =137

$\dfrac{p.2}{17} \cdot \dfrac{137}{5} = \dfrac{274p}{85}$

9. A
$x^2 + 4x - 8 = 4$
$x^2 + 4x - 12 = 0$
$(x+6)(x-2) = 0$
$x=-6 \quad x=2$

10. C
If $y = \dfrac{x^2}{3} - \ln(2x)$ is graphed, its x intercept will be seen to be located at 2.06 and 0.55.

11. A
$x^2 + y^2 + 6x - 8y + 25 = 0$
$x^2 + 6x + 9 + y^2 - 8y + 16 = 0$
$(x+3)^2 + (y-4)^2 = 0$ ➔ $x=-3$ and $y=4$ ➔ The point $(-3, 4)$

12. C
$\cos\theta = \dfrac{3}{5} = \dfrac{4}{6} \Rightarrow h = 10$

$x^2 + 6^2 = 4^2$ ➔ $x=8$ ➔ Area $= \dfrac{6.8}{2} = 24$

13. C
$y=mx+b$
$y=5x+b$
$17=10+b$ ➔ $b=7$

14. E
Let $x=-1$ ➔ $1+b+9 =c$
➔ $c=16$

15. C
$g^{-1}(x) = \dfrac{x+1}{2}$

$f \circ g \circ g^{-1} = f \Rightarrow f(x) = 3x+4 \quad 0 \quad \dfrac{x+1}{2} = 3\left(\dfrac{x+1}{2}\right) + 4$

$= \dfrac{3x+3+8}{2} = \dfrac{3x+11}{2}$

16. D
$9-x^2=0$ so $x^2=9$
$x=\pm 3$ are the values for which y as not defined.

17. C
$x^3 - 5x^2 - 6x = 0$ so $x(x^2 - 5x - 6)=0$
$x(x-6)(x+1)=0$ ➔ $x=0$ or $x=6$ or $x=-1$

18. D
$\sin x\left(3\sin x + \dfrac{2}{\sin x}\right) + 3\cos^2 x$

$= 3\sin^2 x + 2 + 3\cos^2 x = 3(\sin^2 x + \cos^2 x) + 2 = 3+2 = 5$

19. A
$\sqrt{y^2 - 9} \Rightarrow \tan\alpha = \dfrac{\sqrt{y^2-9}}{3}$

20. D
Area of the hexogen is twice the area of the shaded equilateral triangle. So area is $2.\dfrac{12^2\sqrt{3}}{4}.\dfrac{1}{12^2} = \dfrac{\sqrt{3}}{2}$ ft²

21. B
$3^2+(2-1)^2 > 9$ ➔ answer is B.

22. A
$\tan A\hat{B}C = \dfrac{x}{x\sqrt{2}} = \dfrac{1}{\sqrt{2}} \Rightarrow A\hat{B}C = 35.3°$

23. A
$(x+4)^2=x^2+(x+2)^2$
$x^2+8x+16=x^2+x^2+4x+4$
$x^2-4x-12=0$
$(x-6)(x+2)=0$ ➔ $x=6$
Area $= \dfrac{AB.BC}{2} = \dfrac{6.8}{2} = 24$

24. B
$\dfrac{b+c}{a+b+2(b+c)} = \dfrac{4}{2+2.4} = \dfrac{4}{10} = 0.4$

25. C
Shift f(x) towards left for 1 with and you get C.

26. C
Probability that the arrow will land on. Region 2 at least once = 1-prob that it will not land on region 2 in both spins.
➔ $P = 1-\left(\dfrac{3}{4}\right)^2 = 1-\dfrac{9}{16} = \dfrac{7}{16}$

27. B
$\Delta=3^2-4k \geq 0$
$4k \leq 9$
$k < 2.25$ ➔ 2 out of 8 possibilities work ➔ $p = \dfrac{2}{8} = \dfrac{1}{4}$

28. B
$\left.\begin{array}{l} a = e^2 \\ b = 10^{-1} \\ c = 2^3 \end{array}\right\}$ $abc = e^2.\dfrac{1}{10}.8 = \dfrac{4e^2}{5}$

29. D
$\dfrac{1}{3}$ min: n articles ➔ 30 min : 90n articles at the same ratio ➔ 180n articles of twice the given rate.

30. E
$6y^2=3x-18$
$y^2 = \dfrac{x}{2} - 3$ ➔ Infinitely many solutions (x,y) exist.

31. E

$y = x + \dfrac{2}{x}$ and x>0. Graph the function.

$y = x + \dfrac{2}{x}$ for x>0 and find its minimum point.

➜ The point is (1.41, 2.83) ➜ Min y value is 2.83.

32. C

$\overset{\wedge}{1} + \overset{\wedge}{3} = \overset{\wedge}{8}$ may or may not be correct; the rest is always correct.

33. C

$y^6.z^5 = y^5 z^5$ ➜ $y^6 z^5 - y^5 z^5 = 0$
$z^5(y^6 - y^5) = 0$
$y^6 - y^5 = 0$
$y^5(y-1) = 0$ so y=0 or y=1

34. A

Area of the overlap = 10.(6+4-9)= 10

35. E

N has 3x4 positive integer divisors. But it has 24 distinct integer divisors.
Including the negative ones as well.

36. B

$\left.\begin{array}{l} a_5 = 64 \\ a_{11} = 1 \end{array}\right\}$ $\begin{array}{l} a_5.r^6 = a_{11} \\ 64.r^6 = 1 \end{array}$

$\Rightarrow r^6 = \dfrac{1}{64} = 2^{-6} \Rightarrow r = \mp\dfrac{1}{2} \Rightarrow a_3 = \dfrac{a_5}{r^2} = \dfrac{64}{1\!/\!4} = 256$

37. B

Perimeter$= 18\sqrt{3} + \dfrac{18\sqrt{2}}{2} + \sqrt{324 + 162}$

= 65.95 in = 5.5 ft.

38. B

$2\pi R > 100$

$2\pi.4.5t > 100$ ➜ $t > \dfrac{10°}{9\pi} = 3.54$ ➜ minimum integer

value of t=4 sec.

39. A

This question indirectly asks the x intercept of the line that contains (-2, 1) and (-1, 2). Let the line contain

both points. ➜ m=slope=$\dfrac{2-1}{-1+2} = \dfrac{1}{1} = 1$

➜ y=1x+b ➜ 2= -1 + b ➜ b=3 ➜ y=x+3
➜ The x intercept is -3. So answer is A.

40. B

3.-2 = -6
3.8 = 24
7.-2= -14
7.8 = 56
-14 < xy < 56

41. D

Perimeter = $4.4\sqrt{2} = 16\sqrt{2}$

42. D

$\left.\begin{array}{l} \text{center}: (-5,-5) \\ \text{Radius}: 5 \end{array}\right\}(x+5)^2 + (y+5)^2 = 5^2$

➜ Answer is D.

43. B

y=-x+1 and $x \geq 0$ ➜ This is one position of a line.

44. C

If 3+2i is a root then 3-2i is also a root.
Q=(3+2i)(3-2i)=9+4=13

45. C

The amount of pure acid in the solution stays the some.

➜ $60.\dfrac{15}{100} = (60+x).\dfrac{12}{100} \Rightarrow 75 = 60 + x$

x=15

46. D

a+b+c equals the length of a side of $\overset{\Delta}{ABC}$.

47. C

x.(x+7)=14.20
x²+7x-280=0 ➜ x=13.59

48. C

77-7=70 m of the height of the building lies above the trees.

$x = 99 - \dfrac{70}{\tan 55°} - \dfrac{70}{\tan 66} = 18.8m$

49. B

Although B seems to be correct, the graph excludes the people above 64 and below 16 years of age; Therefore B cannot be deduced.

50. C

$1500.\dfrac{4000}{5000}.\dfrac{15}{100} = 180$ people play golf.

Model Test 9 – Solutions

1. C

$\sqrt{56 - x^2} = \sqrt{56 - 7} = \sqrt{49} = 7$ and x²=7

2. B

$\dfrac{3}{8}.100 = 37.5\%$

3. D
$x=6+y^2$
$3(6+y^2)-y=42$
$3y^2-y+18=42$
$3y^2-y-24=0$

➔ $y=\dfrac{-8}{3}$ or y=3

$y-3$ $x=\dfrac{118}{9}$ $x=\dfrac{45}{3}=15$

4. A
2529= 937 + 19.9x
➔ x = 80

5. B
$x^2=9$
$x=\pm 3$
$y(x+1)=15$
if x=3 then y=$\dfrac{15}{4}$
if x = -3 then y= $\dfrac{-15}{2}$

6. A
-1 < x < 3
-2 < x-1 < 2 == |x-1| < 2

7. C
Let a=1 then by trial and error III is even only.

8. E
Infinitely many different lines pass through D, therefore the coordinates of P cannot be determined.

9. E
$x < x^3 < x^2$ is possible if and only if -1 < x < 0.

10. C
$\dfrac{x+x+x}{x.x}=\dfrac{3x}{x^2}=\dfrac{3}{x}$

11.C
7t = 28
t=4

12. D
70 + 6.5 + 3.7

13. A
AC is an increasing line with positive slope. Therefore its reflection A'C' is a d3ecreasing line with negative slope.

14. D
$\dfrac{d}{2}$ equals one side of the square. Therefore area of the square is $\dfrac{a^2}{4}$

15. C
(x,y) = (2,-1) and (x,y)=(6,1) both satisfy the quation
$y=\dfrac{1}{2}x-2$

16. D
x=1-a ➔ y=x-2=1-a-2=-a-1
$xy=(1-a)(-a-1)=-a-1+a^2+a=a^2-1$

17. D
$5^{x-2}=\dfrac{1}{125}=x^{-3}\Rightarrow x-2=-3$
x=-1

18. A
$(n-g).\dfrac{85}{100}$ is the number of boys under 17 years of age.

19. E
$f(x)=3x^2-4x+5$
f(1)=3-a+5=4
-a = - 4 so a=4
f(-1)=3+4+5=12

20. B
distance traveled by the buy is
(-3-(-5)) + (13—5)=2+18=20

21. C
$\dfrac{3}{5}=\dfrac{x}{x+y}$
5x=3x+3y➔2x=3y
y=$\dfrac{2x}{3}$
$AD=x+x+\dfrac{2x}{3}=16$
$\dfrac{8x}{3}=16\Rightarrow x=6$

22. A
When x=0 y1 < 0; y_2=0; y_3, y_4, y_5 > 0

23. D
3.3 = x.1
x=9 ➔ AB=x+1 = 10

24. A
$\left.\begin{array}{l} s=180-A \\ c=90-A \end{array}\right\}s-c=180-A-(90-A)$
= 180-A-90+A
=90

25. C
The difference is 1.5 x 100.000 in the year 2000 and it is the least.

26. E

The total member of calls is (7.8+5.8)x100.000 in the year 2002.

27. B

7-24-25 is a right triangle.

$\sin\theta = \dfrac{24}{25}$

28. D

For D; mode, median and mean all equal 4.

29. C

$\dfrac{12}{36} = \dfrac{x}{45}$

➔ x=15

30. C

$PL = LD = \dfrac{a}{\sqrt{2}}$

$LR = \dfrac{a}{\sqrt{2}} \cdot \dfrac{1}{\sqrt{2}} = \dfrac{a}{2}$

$PR = PL - LR = \dfrac{a}{\sqrt{2}} - \dfrac{a}{2}$

31. C

$\dfrac{50000 - 5000}{15} = \dfrac{5000 - x}{5}$

$\Rightarrow \dfrac{45000}{15} = \dfrac{50000 - x}{x} \Rightarrow x = 35000$

32. E

ac=5

bd=7

ac + bd = 12

33. D

21.7=147 minutes later it blinks fort he 22nd time. 147 minutes is 2 hours and 27 minutes. So at 1.39 pm it blinks. For the 22nd time.

34. B

$\dfrac{x}{-3} + \dfrac{y}{-3} = 1 \Rightarrow x + y = -3$ and y=-x-3.

35. D

A quadratic function is continuous. Since it is decreasing for -5 < x < -1, D must be correct. We cannot be sure of the rest of the answer choices.

36. D

The endpoints are $\left(2 \mp \sqrt{2}, -1 \mp \sqrt{2}\right)$

37. B

$6s^2 = 54$ ➔ $s^2 = 9$ and s=3

$3\sqrt{2} = 2R \Rightarrow R = \dfrac{3\sqrt{2}}{2} = 2.12$

38. B

$f(g(2x)) = 2x$ ➔ $g(x) = f^{-1}(x) = 2x+1$

39. D

$\cos 53° = \dfrac{6}{x} \Rightarrow x = \dfrac{6}{\cos 53°} = 10$

40. B

$(xy)^{\frac{3}{4}} = 2y^{\frac{1}{4}} \Rightarrow$ Take 4th power of both sides $^3y^3 = 16y$

➔ $x^3y^2 = 16$

$(xy)^{\frac{1}{4}} = 1.2x^{\frac{1}{4}} \Rightarrow xy = 1.2^4 \cdot x \Rightarrow y = 1.2^4 = 2.0736$

$\Rightarrow x = \sqrt[3]{\dfrac{16}{4^2}} = 1.55$

41. A

$14\sin^2 x + 14\cos^2 x = 14$

42. C

From triangle similarity

$\dfrac{5x+1}{7} = \dfrac{4x-1}{5} \Rightarrow 25x + 5 = 28x - 7$

12=3x so x=4

43. D

$x = k \cdot \dfrac{1}{y^2} \Rightarrow xy^2 = k$ where k is a constant.

$4.3^2 = x.2^2$

$\Rightarrow x = \dfrac{4.9}{4} = 9$

44. C

Segment AC definitely intersect line m. So m is not parallel to AC.

45. D

$3 - x^2 > 0$

$x^2 < 3$

$|x| < \sqrt{3} \Rightarrow -\sqrt{3} < x < \sqrt{3}$

46. E

3 regions 4 regions 5 regions

47. C

The unitary numbers constitute an arithmetic sequence with a common difference of 9. therefore the 57th term is 775+56.9=1279

48. E

All of the given six shapes/figures can be obtained.

49. B

R=24in/sec = 2ft/sec

Beginning of the 3rd second is the end of the 2nd second.

$2\pi.5R - 2\pi.2R = 2\pi.3R = 6\pi R = 6\pi.2 = 12\pi$

50. B

$$\frac{4}{13} = \frac{x}{9} \Rightarrow x = \frac{36}{13}$$

Shaded areas is $4^2 - 4 \cdot \frac{x}{2} + 9 \cdot \frac{9-x}{2} = 38.5$

Model Test 10 – Solutions

1. B

1728	2
864	2
432	2
216	2
108	2
54	2
27	2
9	2
3	2

$x^3 \cdot y^2 = 1728 = 3^3 \cdot 8^2$
$x + y = 3 + 8 = 11$

2. B

$c = \frac{3a}{4} \Rightarrow \frac{a}{c} = \frac{4}{3}$

3. B
$f(A) = A^9 \rightarrow f(2.1) = 2.1^9 = 794.28$

4. C
$t = 0 \rightarrow h = 100 + 20.0 - 5.0 = 100$

5. B
When (-3, -7) is reflected across the y axis it becomes (3, -7)

6. E
$3.4 \times 10^{-6} = k = -6$

7. C
Sum of the digits must be a multiple of q therefore $29 + E = 9k \rightarrow E$ must be 7.

8. E
Number of distinct integer divisors of N = 4.5 = 20

9. D
When $\alpha + \beta = 90°$, $Sin\alpha = Cos\beta$.

10. A
All choices but A can be rational.
A is definitely irrational.

11. D

$x = \frac{-2y}{y-z} \Rightarrow x - y - xz = -2y$

$xy + 2y = xz \Rightarrow y(x+2) = xz$

$y = \frac{xz}{x+2}$

12. E

$\frac{1}{k(k+1)} = \frac{A}{\underset{(k+1)}{k}} - \frac{B}{\underset{(k)}{k+1}} \Rightarrow I = A(k+1) - Bk$

$\left.\begin{array}{l} k = 0 \Rightarrow A = 1 \\ k = -1 \Rightarrow B = 1 \end{array}\right\} A + 3B = 1 + 3 = 4$

13. E

$m\hat{ABC} = 100° - 30° = 70°$

$\tan \hat{ABC} = $ slope of $AB = \tan 70° = 2.75$

14. B
$y = -2x + b \rightarrow m = -2$

$\rightarrow$ lime m is $y = \frac{-1}{-2}x + b$

$y = \frac{x}{2} + b \Rightarrow 0 = 0 + b$ and $b = 0 \Rightarrow y = \frac{x}{2}$

$r - 1 = \frac{r}{2}$, so $\frac{r}{2} = 1$ and $r = 2$

15. C
The graph of $y = (x^2-1)(x^2-4)-19$ intersect the x axis twice. So the equation has two real solutions.

16. E
$(3+1)^2 + (2+1)^2 = 25$
$16 + 9 = 25$
$25 = 25 \rightarrow$ Circle in choice E satisfies the given condition.

17. B
Area of the rhombus is

$2\left(\frac{1}{2} \cdot 15 \cdot 15 \cdot Sin70°\right) = 211.43$

18. A

$\tan\alpha = \frac{5}{12}$

19. B
$1 - Sin^2x = 0.4$
$Cos^2x = 0.4$

20. D

Shaded area $= \frac{\pi \cdot (2.5)^2}{360°} \cdot 40° = 2.18$

21. B
f(x) is reflected across the y axis, then translated 1 unit rightwards.
So $g(x) = f(-(x-1)) = f(-x+1)$

22. B

$\tan\alpha = \frac{3}{\sqrt{y^2 - 9}}$

23. D
The required point is the midpoint of the segment whose endpoints are the given points. So, the point is
$$\left(\frac{1+5}{2}, \frac{2-10}{2}\right) = (3,-4)$$

24. A
y takes only the values -2, 2 and 3.

25. C
When x is 0, y is minimum and it equals,
$y = 1 + \sqrt{1+0} = 2$.
So the range of f(x) is y ≥ 2.

26. A
$$t = \frac{x+3}{2} = \frac{y-1}{-3} \Rightarrow 2y - 2 = -3x + (-9)$$
2y = -3x-7 ➔ $y = \dfrac{-3x-7}{2} \Rightarrow$ slope is $\dfrac{-3}{2}$

27. C
AB equals 2r end thus
$2r = \sqrt{5^2 + 12^2} = 13$
r = 6.5

28. D
All can be obtained except fort he hyperbola.

29. C
5 4 3 2 1 4 ➔ 480 ways are possible.

30. E
Inverse of f(x)= $\dfrac{x-1}{2}$ is f⁻¹(x)=2x+1 and it does not equal f(x)

31. B
$6a^2 = 24$
$a^2 = 4$
a= 2
Required distance = half the length of the diagonal
$= \dfrac{2\sqrt{3}}{2} = \sqrt{3}$

32. C

There are three such triangles.

33. E
$$\binom{5}{0} + \binom{5}{1} + \binom{5}{2} + \binom{5}{3} + \binom{5}{4} + \binom{5}{5} = 2^5 = 32$$

34. A
$(x+4)^3 = x^3 + 316$ ➔ $x^3 + 12x^2 + 48x + 64 = x^3 + 316$
$12x^2 + 48x - 252 = 0$
$3x^2 + 12x - 63 = 0$ ➔ x=3 OR x=-7

35. D
(a-b)(a+b) = 75
Since a, and b are both positive, a-b < a+b
1 . 75
3 . 25
5 . 15
Since a,b are both positive, a-b < a+b

36. B
Since coefficients are rational, irrational roots must appear in conjugate pairs. Thus B is not possible.

37. B
$x^2 = 7 . (7 + 4) = 77$
$x = \sqrt{77} = 8.77$

38. A
There are 8 x 7 x 6 x 4 = 1344 such numbers.

39. C
$$\frac{x.(x-1)}{2} = 28 \Rightarrow x^2 - x = 56$$
$x^2 - x - 56 = 0$
(x-8)(x+7) = 0 so x= 8
$8 = \dfrac{10}{100}.n$
n= 80

40. D
$\dfrac{x}{y}$ is negative thus $\dfrac{3x}{y}$ is less than $\dfrac{x}{y}$

41. E
y > 0 ➔ (x-2)(x+1) > 0 ➔ x > 2 OR x < -1

42. E
$x = \dfrac{\sqrt{2}}{2}$ and $y = \dfrac{\sqrt{2}}{2}$ ➔ all relations are satisfied.

43. D
The points equidistant from a point and a line that does not contain the given points make up a parabola.

44. C
1 + 5d = 10 ➔ 5d = 9
d=1.8
➔ 1,2.8,4.6,6.4,8.2,10 are the terms.

45. E
|x-1| is less than or equal to 8 but it cannot be zero. So answer is E.

46. A
The resulting object is one fourth of sphere whose radius is 87.
➔ $\dfrac{1}{4}.\dfrac{4}{3}.\pi.7^3 = \dfrac{\pi.7^3}{3} = 359$

47. C

Circle has infinitely many lines of symmetry. Any line that passes through the center of the circle is a line of symmetry.

48. B

c= an + b

$\left.\begin{array}{l} 2000 = 800a + b \\ 3200 = 1400a + b \end{array}\right\}$ a = 2 and b = 400

➔ c = 2n + 400

when n=500 c=2.500+400 = 1400

49. C

$\log_3 \dfrac{x+5}{x-5} = 2 \Rightarrow \dfrac{x+5}{x-5} = 9$

x+5=9x ➔ 50 = 8x

x= 6.25

50. B

The shaded region represents the points on or above the curve y= |x-4|. Therefore y >|x-4|.

Model Test 11 – Solutions

1. E

2a − 2b + 1= 2(a-b)+1 = even + odd = odd

2. C

5 . 4² = 5.16 = 80

3. D

Such lines have unequal slopes. So II and III are the only possible choices.

4. D

$\left.\begin{array}{l} x^2 + y^2 = 5 \\ x^2 - y^2 = 3 \end{array}\right\}$ $2x^2 = 8$ $x^2 = 4$

5. C

$\dfrac{5}{6}x = 0 \Rightarrow x = 0$

$\dfrac{6}{5} - x = \dfrac{6}{5} - 0 = \dfrac{6}{5}$

6. B

$x\left(\dfrac{1}{y} - \dfrac{1}{z}\right) = x\left(\dfrac{z-y}{yz}\right) = \dfrac{xz - xy}{yz}$

7. B

n= 13

m=1-15=-14

8. E

Sum of the consecutive integers equals median times the number of terms and it is - 4 x 10 = - 40

9. D

$\left.\begin{array}{l} xy = 35 \\ yz = 45 \end{array}\right.$ $\left.\begin{array}{l} y = 5 \\ x = 7 \\ z = 8 \end{array}\right\}$ x + y + z = 21

10. A

$x^{odd} < 0$ ➔ x must be negative

11. D

$2^c - c^2 = 1$ ➔ c=1 satisfies this equation.

12. B

$\dfrac{n(n-3)}{2} = 9 \Rightarrow n^2 - 3n = 18$

n² - 3n − 18 = 0 ➔ (n-6)(n+3) = 0

n=6

13. A

$t = -1 \Rightarrow z = (-1)^3 - 1 = -1-1 = -2$

$\left.\begin{array}{l} y = 3(-1)^2 + 4 = 3 + 4 = 7 \end{array}\right\}$ y − z = 7 − −2 = 9

14. C

15. D

$\dfrac{x}{AC} = Sin\theta \Rightarrow AC = \dfrac{x}{Sin\theta}$

16. D

$f(g(2)) = f(2^3 + 2 + 1) = f(11) = \sqrt{11-10} = \sqrt{1} = 1$

17. C

$I + 6 = \dfrac{4}{5}(s + 6)$

$I + 15 = \dfrac{7}{8}(s + 15)$

➔ I = 6 and s = 9

18. D

I and II are correct. III may be false (think of a trapezoid).

19. B

There has to be a that is not a Z since the elements that are both M and T are not Z.

20. C

Back-solving works best fort his question.

1st store : $\dfrac{128}{2} + 4 = 68\$$ spent, 60$ left. 3rd store

$\dfrac{20}{2} = 10\$$ spent.

2nd store: $\dfrac{60}{2} + 10 = 40\$$ spent, 20$ left. 10$ left for hot candies.

21. C

$$d_A = \sqrt{(1-1)^2 + (-1-7)^2} = 8$$
$$d_C = \sqrt{(6-1)^2 + (-1-t^2)^2} = \sqrt{26}$$
$$d_B = \sqrt{(1-8)^2 + (-1-(-1))^2} = 7$$
$$d_D = \sqrt{(1-2)^2 + (-1-7)^2} = \sqrt{65}$$
$$d_E = \sqrt{(1-3)^2 + (-1-5)^2} = \sqrt{40}$$

d_C is minimum

22. C
A & D are correct for every angle.
B and E are correct in this particular case.

C is false because $\tan\theta = \dfrac{12}{5}$

23. C
f(f(A))=f(A²-9)=(A²-9)²-9=0
A² - 9 = ±3 ➔ A² = 12 or 6 ➔ A = ±3.5 or ± 2.4
so A cannot be 0

24. A
x² < x ➔ x² - x < 0
x (x-1) < 0 ==< 0 < x < 1

25. D

Area of shaded region = $\pi.\left(\dfrac{17}{2}\right)^2 - 15.8 = 107$

26. B
$\dfrac{x}{y} = \dfrac{3}{8}$

x + y = -44
x= 3t
y=8t ➔ x + y = 11t = -44
t= -4 ➔ x= -12 and y= -32
Lesser is -32 and B is correct.

27. C
Except for the letter in C, all other letters have one line of symmetry; but the letter in C has two lines of symmetry.

28. C
C is a subset of all other intervals.

29. B
2b = 2a + x   and b=a+y
➔ 2b – 2a = x   and b-a=y
➔ 2y = x ➔ $y = \dfrac{x}{2}$

30. E
x + 1 ≥ 0 and x-1 ≠ 0
➔ x ≥ -1 and x ≠ 1

31. D
$$u = \frac{13+14+15}{2} = 21$$
$$s = \sqrt{21.8.7.6} = 84$$

32. E
$$x + \frac{x}{\sqrt{2}}.2 = 100$$

$$x + x\sqrt{2} = 100 \Rightarrow x = \frac{100}{\sqrt{2}+1}.\frac{\sqrt{2}-1}{\sqrt{2}+1} = 100\left(\sqrt{2}-1\right) \text{Ans}$$

wer is neither of the choices A, B, C and D.

33. B
The resulting figure can also be obtained by a single reflection in the line y = -x.

34. C
$$2\pi R + 4R = 100 \Rightarrow R = \frac{100}{2\pi+4} = 9.725$$
$$4R^2 + \pi R^2 = \left(4+\pi\right).R^2 = 675.4$$

35. B
For II, g(-x) = f(x) – f(-x) = -(f(-x)-f(x))=-g(x).

36. B
$$\frac{A}{9} \quad \frac{B}{10} \quad \frac{C}{10} \quad \frac{B}{1} \quad \frac{A}{1} = 900 \qquad \text{possibilities}$$

37. C
26³.10²-26².10³=1081600 which is closest to 1100000

38. B
Cosθ cannot equal 1 since θ cannot be 0°
Cosθ cannot equal -1 since θ cannot be 180°

39. C
$$\pi.4^2.6 + \frac{\pi.4^2.4}{3} = 368.6 \approx 369$$

40. D
$$10t - 1 = \frac{4t + 12t + 2}{2}$$
20t – 2 = 16t + 2 ➔ 4t = 4
T=1 == 4,9,14,…
A₅₀=a+49d = 4 + 49.5 = 249

41. C
When center is (h₁k) and radius is R1(x-4)² + (y-k)² = R² is the equation of the circle.
(x-3)² + (y—1)² = 10² ➔ (x-3)² + (y+1)² = 100

42. D
x+1=0
x= -1
P(-1+2)=P(1) is the remainder.

43. D

$$M\left(\frac{1+3}{2},\frac{2+6}{2}\right)=(2,4)$$

$$\text{slope}=\frac{-1}{\frac{6-2}{3-1}}=\frac{-1}{2}$$

$$y-4=\frac{-1}{2}(x-2)$$

44. C

$$(2\pi.3+2\pi.5).\frac{72°}{360°}+2.2=14.05$$

45. A

distance from (3,5) to the line $3x-4y+1=0$ is

$$d=\frac{|3.3-4.5+1|}{\sqrt{3^2+4^2}}=\frac{|9-20+1|}{5}=\frac{10}{5}=2$$

46. B

$x = 10. \sin20°.2=20\sin20°=6.84$

Shaded area $= 2.\dfrac{1}{2}.10.10.\sin40°$

$$\rightarrow \frac{6.84^2}{2}=40.89\approx41$$

47. A

$\tan^{-1}1=45$ and $\tan^{-1}\sqrt{3}=60°$

➔ Acute angle between the links is 60° - 45° = 15°

48. A

$$P(\text{T or B})=\frac{10+7+13}{42}=\frac{30}{42}=5/7$$

49. E

Number of such patterns is s! = 120

50. B

f(x) = 2(x²+6x)+3= 2(x²+6x+9)-15= 2(x+3)³-15

f(x-3) is symmetric about the y axis ➔ k=3

Model Test 12 – Solutions

1. C

a.1+4 = 3.2 ➔ a=6-4=2

2. C

$p^5.r^7$ is even because it is the product of an odd and an ever number.

3. C

10.3+8(n-3)

10: cost per hour fort he first 3 hours

3: First three hours

8: Cost per hour thereafter

n-3: resulting number of hours.

4. C

|-2.8| - |5.4| + |-0.6| = 2.8-5.4 + 0.6 = 3.4 – 5.4 = -2

5. D

2x=5y and (2x+x+2y).2=76

2x=5y and 6x+4y=76

X=2.5y ➔ 5x2.5y+4y=76

$19y = 76$ ➔ y=4 ➔ x=$\dfrac{20}{2}$=10 ➔ Area = 20x18=360

6. A

$$\frac{n.50}{60}+\frac{m.60}{60}=\frac{5n}{6}+m$$

7. B

$\dfrac{a}{b}$ =p and p is a positive integer, so I is true.

II is not necessarily true.

$\dfrac{d}{c}=\dfrac{1}{r}$ and $\dfrac{1}{r}$ is less than I so III is not true.

8. B

Shaded area $=\dfrac{\pi.4^2}{2}-\dfrac{4.4\sqrt{3}}{2}=11.27\approx11.3$

9. E

180-b+a+c = 180

➔ b=a+c

10. B

One angle of the pentagon is $\dfrac{3\times180}{5}=108°$

$m\hat{CBG}=108°-90°=18°\Rightarrow x=\dfrac{180-18}{2}=\dfrac{162}{2}=81°$

11. D

$\log_381 < \log_384 < \log_3243$

$4 < \log_384 < 5$

12. E

Fort he y intercept, x =0 ➔ t=2 and y=2+2 = 4

13. B

6x6=36 so D and E whose scores are 39 and 41 must have received 7 from one or more of the referees.

14. A

product of the slopes of the lines must be -1.

$3.\dfrac{2}{k}=-1\Rightarrow k=-6$

15.A

$\dfrac{x-1}{x}\geq1\Rightarrow\dfrac{x-1}{x}-1\geq0\Rightarrow\dfrac{-1}{x}\geq0=\dfrac{1}{x}\leq0\Rightarrow x<0$

16. B

a= 2R = 2x0.42 = 0.84

Area = a²=0.84² = 0.7056 = 0.71

17. A

–f(-x) is equivalent to reflecting the graph in the y axis then the resulting graph in thr x axis. This is equivalent to a reflection in the origin.

18. D

$$\frac{(x+3)(x-1)}{x-2} \cdot \frac{(x-2)(x+1)}{x+3} = x^2 - 1$$

19. D

Coordinates A are $\left(-4, \frac{-4\sqrt{3}}{2}\right) = \left(-4, -2\sqrt{3}\right)$

20. A

Coordinates of another vertex can be:

$(\mp 6, 0)$; $(0, \mp 6)$; $(\mp 3, \mp 3)$; $(\mp 3, \pm 3)$; $(\mp 3\sqrt{2}, 0)$; $(0, \mp 3\sqrt{2})$

21. E

$\left. \begin{array}{l} a_4 = n = a + 3d \\ a_n = 4a + (n-1).d \end{array} \right\}$ $\begin{array}{l} a + 3d = n \\ a + (n-1)d = 4 \end{array}$

$(4-n)d = n-4$ ➔ $d = -1$

➔ e=n43 ➔ $a_2 = e + d = n + 3 - 1 = n + 2$.

22. A

5=e

8 = b+d+f

a+b+c+d+e+f+g=18 ➔ e4c+g=18-5-8=5

23. A

$m D\hat{C} E = \frac{180 - 150}{2} = \frac{30}{2} = 15°$

$m A\hat{F} C = 90° + 15° = 105°$

24. C

2 (6k + 10k) =64 so k = 2

and the total area is $\frac{6.8}{2} \cdot 8 = 24.8 = 152$

25. D

$\tan B = \frac{5}{6}$

26. E

4-x² ≥ 0 ➔ -x² ≥ -4

x² ≤ 4

-2 < x < 2 ➔ 3 is not in the domain.

27. B

The required region is below the line y=-2x+1 and between the lines x=-1 and x=2, x=2 inclusive. Correct answer is B.

28. C

I. f(0) = f(2) = 1 is correct

II. $f(-1) = f(3) = \frac{1}{2^2} = \frac{1}{4}$ is correct.

III is not correct.

29. D

f(x) = x-2

g(x)=x42 when x≠2

30. E

AB < OA + OB

➔ AB < OR + OM ➔ I is correct

AB < BR + RA

➔ AB < BR + BM ➔ II is also correct. Since AR = RM, OR bisects style $A\hat{O}M$.

31. B

x²+8² = (16-x)² = 256 -32x + x²

x² + 64 = 256 – 32x + x²

32x = 192

x = 6

32. B

7 – x + 8 – x = 10

15-2x =10

2x=5

$x = \frac{5}{2}$

33. D

There are 27 smaller cubes and 3x3x6=54 exposed faces each of which is a square. To maximize the yellow surface we place the yellow cubes in the corners and in the middle of each side so we will obtain a total of 8x2 + 4x2 = 32 yellow and 54-32=22 blue faces.

$\Rightarrow \frac{32}{22} = \frac{16}{11}$

34. E

Radius is 3 ➔ After the rotation the resulting solid is a full sphere 8not a half sphere)

$\Rightarrow \frac{4}{3} \cdot \pi \cdot 3^3 = 4.9\pi = 36\pi$

35. C

$\sqrt{n^3 - 4} = AA$ where AA is a two digit integer.

Back solving is the best way in solving this problem.

$\sqrt{5^3 - 4} = \sqrt{125 - 4} = \sqrt{121} = 11$ and 11 is a palindrome.

36. B

a-4=b

B= 20b=2a

➔ a-4 = 2a ➔ a=-4

B=2.(-4)=-8

37. B
Line d is perpendicular to the given line segment.
Therefore slope of d is

$$\frac{1}{\frac{3-0}{0--3}} = \frac{-1}{1} = -1$$

38. E
Correct answer is given in E where n is the number of days that passed.

39. E
I f(x) I reflect the negative y portion with respect to the x axis. Therefore E can be the graph of f(x) before the reflection.

40. C
$\alpha = 90°$ and $\theta = 90° - 40° = 50°$
Therefore C is the correct answer.

41. C

$$d = \frac{x}{2}\sqrt{3} \cdot 2 = x\sqrt{3} \Rightarrow \frac{d}{x} = \sqrt{3}$$

42. B

$$\frac{PR}{SinS} = 2R \text{ by the sine rule.}$$

$$\frac{10}{Sin60°} = 2R \Rightarrow R = \frac{5}{\frac{\sqrt{3}}{2}} = \frac{10}{\sqrt{3}} = 5.8$$

43. E

$$\log_8 3 = \log_{(2^3)}3 = \frac{1}{3}\log_2 3 = \frac{1}{3} \cdot \frac{1}{x} = \frac{1}{3x}$$

44. A
In triangle ACA' $1 < 2x < 13$
$0.5 < x < 6.5$ → x cannot be 6.5

45. B
$2(3.5 + 3.n + 5.n) < 2(2.3 + 2.2n + 3.2n)$
$30 + 16n < 12 + 20n$
$18 < 4n$ → n> 4.5 → min(n) = 5

46. C

$$n = 6 - \sqrt{2} \Rightarrow Area = \frac{(5\sqrt{2}+2)(6-\sqrt{2})}{2} = 20.8$$

47. D
Square is both a rhombus and a rectangle.
So correct answer is D.

48. E
Number of distinct positive divisors of N is
$(141)(1+4) = 2.5 = 10$

49. C
Points A and E are inside the circle since they make
$(x-1)^2 + (y+1)^2$ less than 9.

50. E
The points make up a trapezoid and the points inside the trapezoid.

Model Test 13 – Solutions

1. D
A,B,C,D are equivalent to $M = -n^6$
E is equivalent to $m = n^6$.

2. E
Since $0 < y < 1$ and $y^x < y$, $x > 1$

3. D
$\sqrt{2}x + 2y = 1$ and $x - \sqrt{2}y = \sqrt{2}$ so

$$2\sqrt{2}x = 3 \Rightarrow x = \frac{3}{2\sqrt{2}} \cdot \frac{\sqrt{2}}{\sqrt{2}} = \frac{3\sqrt{2}}{4}$$

4. A
$m+n = 2p$

5. D
the graph of the line lies in the quadrants 1,2 and 3.

6. E
$\left.\begin{matrix} pa + qb = 7 \\ pb + qa = 8 \end{matrix}\right\}$ $\begin{matrix} pa + pb + qa + qb = 15 \\ p(a+b) + q(a+b) = 15 \end{matrix}$
$(a+b)(p+q) = 15$ → $a+b = 5$ → $a^2 + 2ab + b^2 = 25$

7. C
Area = 3.5 = 15

8. D

$\frac{\pi}{2}$ is not rational since π is not rational.

9. A
$\begin{matrix} a+1+b-3 = 0 \\ a-2-b-3-5 = 0 \end{matrix}$ $\begin{matrix} a+b = 2 \\ a-b = 10 \end{matrix}$
2a=12 → a=6 → b= - 4 → ab= -24

10. C

Ad	St
x	900-x	3x+2(900-x)=2300
3	2	1800+x=2300
		x=500

11. D

$$\frac{2(Sin^2x + Cos^2x)}{Cosx} = 5 \Rightarrow \frac{2}{5} = Cosx \Rightarrow x = Cos^{-1}\frac{2}{5} = 66°$$

12. E
In E, mean = 103.2 and median = 103.

13. A
The correct answer is given in A; customer should

$$pay \; c \cdot \left(1 + \frac{p}{100}\right) \cdot \left(1 - \frac{d}{100}\right) \cdot \left(1 + \frac{t}{100}\right).$$

14. A
Correct translation to mathematics is given in A.

15. D
D is the smallest.

16. C
If 3+4i is a root, then 3-4i is also a root.
Then P=344i + 3-4i = 6

17. B
$\dfrac{225°}{360°} = \dfrac{x}{100} \Rightarrow x = 62.5\%$

18. C
The resulting solid is a cone with R=3 and h=7
➔ $v = \dfrac{\pi.3^2.7}{3} = 21\pi = 65.97$

19. C
$A = \dfrac{1}{2}.\dfrac{1}{3} = \dfrac{1}{6}$

$B = \dfrac{1}{2}.\dfrac{1}{4} = \dfrac{1}{8}$

$\dfrac{1}{6} + \dfrac{1}{8} = \dfrac{7}{24}$

20. C
x coordinate of the circle must be equal to the radius of the circle.
➔ $(x-4)^2 + (y+6)^2 = 4^2$ ==< center (4, -6); R=4

21. A
The vertex is at (2,120), thus the object reaches the maximum height of 120 meters when t=2.

22. B
$(m-o)^2+(m-o)^2=5^2$ ➔ $(4m-o)^2(4n-o)^2=(4.5)^2$ ➔ d=20

23. B
P=1,   Q=1   P=5,   Q=3^4
P=2,   Q=3   P=6,   Q=3^5
P=3,   Q=3^2 P=7,   Q=3^6=729
P=4,   Q=3^3 P=8,   Q=3^7 > 2187 ➔ p=8 is
printed.

24. B
When the curve is reflected across the x axis it becomes $y=-(x^3-x^2)=-x^3+x^2$

25. B
B is the correct answer

26. C
d: |-2+2i-(6-4i)| = |-2+2i-6+4i| = |-8+6i| = 10

27. E
1.5mi = 1.5 x 1760 = 2640yd and 15ft = 5yd
$\tan\theta = \dfrac{h-5}{2640} = \tan 2 \Rightarrow h = 97.19yd.$

28. B
$prob = \dfrac{1}{5}.(1-0.35) = \dfrac{0.65}{5} = 0.13$

29. C
$\dfrac{x}{P} + \dfrac{y}{-Q} = 1$ so $\dfrac{x}{P} - \dfrac{y}{Q} = 1$

30. D
$5^2+h^2=18^2$ ➔ h=17.29 ➔ Area = 86.5

31. B
$\tan 40° = \dfrac{21}{7+x}$ ➔ $x = \dfrac{21}{\tan 40°}-7 = 18.03$

32. A
Sophomores: x: 80°
Seniors:y:120° 120°-80°=40°
Freshmen: 360°-(80+120+100)=60°
$\left.\begin{array}{l}40° \to 24\\ 60° \to x\end{array}\right\} x = \dfrac{60.24}{40} = 36$

33. C
Area of ABCD=$\dfrac{18.18}{2} = 162$ in $\dfrac{162}{144} = 1.125ft^2$

34. C
The unitary numbers are all divisible by 9.
➔ 108, 117, 126,..., 999 ➔ $\dfrac{999-108}{9}+1=100$
unitary numbers have three digits.

35. A
Line m: Slope=$\dfrac{6-1}{3--2} = \dfrac{5}{5} = 1$ ➔ y= 1x+b
6=3+b ➔ b=3 ➔ y=x+3   y=x+3
y=-x+3
x+3=-x+3 so x=0 and y=3
so (0, 3)
Line n : y-1=-1(x-2) ➔ y= -x+3

36. B
f(x-1)=3(x-1)+3=3x Correct answer is given B.

37. C
y=x² -1
x=y²-1 ➔ y² = x+1
$y = \mp\sqrt{x+1} = f^{-1}(x)$

38. A

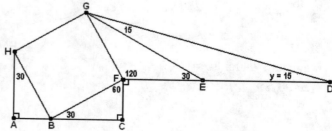

39. A

C: (-5, 0) ➔ |-5+5| + |0-5| +k=0

5+k=0

k=-5

40. E

(x,y)=(-2x,y)

x=-2x ➔ 3x=0 ➔ x=0 which represents the y axis.

41. A

The number of shaded squares is

$$\frac{n(n+1)}{2}+1=\frac{n^2+n+2}{2}$$

42. B

$$2^{x+1}=0.25=\frac{1}{4}=2^{-2}$$

x+1=-2 ➔ x=-3

$$\log_3(4-2(-3))=\log_3 0=2.096$$

43. B

Shaded area = 24.10 -π.s²=161.46 ≈ 160

44. E

II and III are correct

I is false because if the line belongs to one of the planes, it does not intersect the other.

45. C

f(f(-2) + f(1) + 1)

= f(-1+0+1)=f(0) = 0

46. C

When n=8, 2n+1=17 is prime; 2n-1=15 and 2n+5=21 are not prime.

47. A

$$f^{-1}(x)=\frac{x+1}{2}$$

$$f^{-1}\text{ofug}=q\Rightarrow g(x)=\frac{x+1}{2}.3x+4=\frac{3x+5}{2}$$

48. B

$$C=20000\cdot\left(\frac{3}{5}+\frac{2.12}{7}-\frac{5}{12}\left(\frac{12}{8}\right)^2\right)$$

$$=61821\approx62000$$

49. C

$$\frac{100.211}{350}=\frac{6.5}{h}$$

➔ h=22.7

50. D

Maximum value: f(0)=1+√0 = 1

A ≤ f(x) ≤ 6

Model Test 14 – Solutions

1. E

$$\left[3-10.\frac{-1}{7}\right]^{-1}=\left(3+\frac{10}{7}\right)^{-1}=\frac{7}{31}=0,226$$

2. B

$$x^2=\frac{4}{2}=2.\Rightarrow(2.4+6)/(2.4+5)=\frac{14}{13}$$

3. C

4x²-y²+2x-y = (2x-y)(2x+y)+(2x-y)

        = (2x-y)(2x+y+1)

4. D

2 + 3 + 5 + 7 + 11 + 13 + 17 + 19 = 77

The greatest prime factor of 77 is 11.

5. E

9-x ≥ x-3

-x²-x+12 ≥ 0

x²+x-12 ≤ 0

(x+4)(x-3) ≤ 0

➔ -4 ≤ x ≤ 3

6. B

51+57+63+69+....+495

$$\text{sum}=\left(\frac{495-51}{6}+1\right)\left(\frac{495+51}{2}\right)=20475$$

7. B

$$x\left(1+\frac{9.75}{100}\right)-x.\left(1+\frac{8.50}{100}\right)=15$$

$$x.\frac{1.25}{100}=15\Rightarrow x=1200$$

8. D

x + y + z = 180° ➔ y-z+y+z = 180°

Y= 90° ➔ since one angle is 90°. It is a right triangle.

9. E

In E, f(-x)=-7(-x)³+(-x) = 7x³-x=-f(x)

10. C

When $\frac{y}{x}>1$, $\frac{y}{x}$ is positive. So all such points must lie in quadrants I and III.

11. E

$$x\sqrt{2}=4$$

$$x=\frac{4}{\sqrt{2}}=\frac{4\sqrt{2}}{2}=2\sqrt{2}$$

12. E

a³ ➔ 400

(3a)³ ➔ x

$$\frac{a^3}{(3a)^3}=\frac{400}{x}$$

x=27x400 = 10800

13. C

$$30 \times 100 \times \frac{2.55}{100} = 76.5$$

14. A

Consider Bilal and Hatice as 1 person.
➔ 4 people can sit in 3! Ways at a round table.
But Bilal and Hatice can alternate so 3! . 2 = 12 ways are possible.

15. B

$0 < a < 1$
$-1 < -b < 0$ ➔ $0 < b < 1$
➔ $0 < ab < 1$

16. C

Back-solving works the best fort his problem.

$\frac{64}{2} + 2 = 34$; $64 - 34 = 30$ left for the second store.

$\frac{30}{2} + 5 = 20$; $30-20 = 10$ left fort he third store.

$\frac{10}{2} = 5$ spent at the third store and 5 left for hot chocolates.

17. D

I implies that the quadrilateral is a parallelogram. Each of II and III imply that the quadrilateral is a rectangle.

18. C

B and E are correct for every angle.
Since θ is less than 45°, $\tan\theta < 1$. So c is false.
A and D are both correct.

19. D

If we know the room assignments of the seniors or the room assignments of a senior and a junior then we can determine all room assignments as we know that a senior occupies a room alone.

20. B

The center must be (-R, R) where R is the radius of the circle. Therefore answer is B.

21. D

$$\frac{2}{5}\left(1-\frac{4}{7}\right) = \frac{2}{5}.\frac{3}{7}$$

22. B

The rule is : $a_{n+1}=a_n.2+2$ ➔ $a_7=62.2+2=126$

23. E

Any line passing through the center of the square can portion it into two congruent rectangles.

24. D

D is the superset of all other intervals.

25. E

$2b=x+180-2a$ ➔ $2a++2b=180+x$
$y+a+b=180$ $2(180-y)=180+x$
$a+b=180-y$ $360-2y=180+x$
             $180-2y+x$
             $90=y+0.5x$

26. D

Heron's formula is used.

$$u = \frac{26+28+30}{2} = \frac{84}{2} = 42$$

Area $= \sqrt{u(u-a)(u-b)(u-c)} = \sqrt{42.16.14.12} = 336$

27. E

$ma + nb - na - mb=7-5=2$
$(m-n).a+(m-n)(-b)=2$
$(m-n)(a-b)=2$ ➔ $a-$

$b=\frac{2}{0.4} = 5 \Rightarrow a^2 - 2ab + b^2 = 5^2 = 25$

28. A

$$s_1 = \frac{27.h}{2}$$

$$s_2 = \frac{27.(32-h)}{2}$$

$$s_1 + s_2 = \frac{27.32}{2} = 27 \times 16 \quad 11^2 = \frac{27 \times 16}{9} = 48yd^2$$

29. A

$2\sin(3\theta)=\cos^2(2\theta) + \sin^2(2\theta) = 1$

$\sin(3\theta)=\frac{1}{2} \Rightarrow 3\theta=30°$

           $\theta=10°$

30. C

$5 - 6 - 7$
$5 - 7 - 6$
$6 - 5 - 7$
$6 - 7 - 5$
$7 - 5 - 6$
$7 - 6 - 5$
Six different paths.

31. C

$x^2-5x+p=0 \Rightarrow x^2--2x+3=0$
                   $x^2+2x+3=0$
Any nonzero multiple of x^2+2x+3 will give the designated zeros.

32. D

$t^2=-x$
$y=(-x)^2-1=x^2-1$ and $x \leq 0$
$\Rightarrow$ The parametric equations represent part of a parabola.

33. D

A single reflection about the origin

34. D

$y = AP + PD = \sqrt{7.5^2 + x^2} + \sqrt{4.5^2 + (9-x)^2}$

Maximum value of y is when x=5.625
When x=0
X=0 $\Rightarrow$ y=17.6

35. C

x-3=2 $\Rightarrow$ x=5
x-3=-3 $\Rightarrow$ x=0
$\Rightarrow$ {5,0}

36. D

-2-y=2-x+1
-y=-x+5
y=x-5

37. B

$2\pi R = 5 \Rightarrow R = \dfrac{5}{2\pi}$

$V = \pi.\left(\dfrac{5}{2n}\right)^2 .3 = 5.97$

38. C

$R = \dfrac{3}{2} + \dfrac{1}{2} = 2$

$R_1 = 1+1 = 2 \qquad R_2 = 0+2 = 2 \qquad R_3 = 0.5 + 0.5 = 1$

$R_1 = R_2 = R = 2$

39. A

g(x)=-(x+2)²+1
G(-1.2)=-(0.8)²+1=1-0.64=0.36

40. A

y= 6(-2x+1)=-12x+3 $\Rightarrow$ slope = -12

41. B

I and II do not pass the vertical line test; therefore they do not represent functions.

42. B

Reflection of (x,y) in the y axis gives (-x,y), not (-y,x)

43. B

A: rotation
B: horizontal stretch
C: line reflection
D: point reflection
E: vertical stretch

44. E

P(0) = 3= c
P(-1) = a-b+3=5 $\Rightarrow$ a-b=2

45. D

x²-6x+9+y²+ky+4 = 1+9+4
(x-3)² + (y42)² = 14 = R² $\Rightarrow$ R = $\sqrt{14}$

46. C

$\tan 31.5 = \dfrac{1972}{x}$

$x = \dfrac{1972}{\tan 31.5} = 3218 \Rightarrow$ best answer is C.

47. B

y-17=3(x-5)
y=3x+2
x=0 $\Rightarrow$ y=2

48. E

Infinitely many integers satisfy the given inequality.

49. A

24-15= 9ft.

50. D

Locus of all such points is the minor arc AB of the circle shown. The radius of this circle is given by

$\dfrac{AB}{Sin120°} = 2R$

$\Rightarrow 2R > AB$

Model Test 15 – Solutions

1. D
A & B : Associative property
C & E: Commutative property
A: Distributive property (multiplication of addition)

2. B
AD = 4DC
12+x=4x
12 = 3x $\Rightarrow$ x=4

3. D
Whether n is odd or even, II and III are always even.

4. C
Intersection 5 $\leq$ c < 7

5. D

$a^r = 2.4 \Rightarrow a^{-2r} = \left(a^r\right)^{-2} = 2.4^{-2} = \dfrac{1}{2.4^{-1}} = \dfrac{1}{5.76}$

6. E
x+5 must be positive and x-3 must be nonzero
So x > -5 and x≠3

7. A

$\dfrac{4(M-5)+20}{4} = \dfrac{4M-20+20}{4} = \dfrac{4M}{4} = M$

8. D

$\dfrac{2x-8}{x-8}.\dfrac{x^2-5x-24}{x^2-16} = \dfrac{2(x-4)}{x-8}.\dfrac{(x-8)(x+3)}{(x-4)(x+4)} = \dfrac{2x+6}{x+4}$

9. B

$$s+e=\frac{s-e}{t}\Rightarrow st+et=s-e$$
$$e+et=s-st$$
$$e(1+t)=s(1-t)$$
$$e=\frac{s(1-t)}{1+t}$$

10. B

$$f^{-1}(x)=\frac{3x-4}{12}\Rightarrow f(x)=\frac{12x+4}{3}=\frac{4(3x+1)}{3}$$

11. E

$$A=\frac{k}{\frac{1}{B}}$$
$$A=kB$$
$$8=k.4\Rightarrow k=2$$
$$B=5\Rightarrow A=2.5=10\Rightarrow A^2=100$$

12. E

$$30°\times3+30°\frac{20}{60}=90°+10°=100°$$

13. B

$$x^2=10^2+24^2\Rightarrow x=26$$
$$\sin53°=\frac{26}{2R}\Rightarrow 2R=\frac{26}{0.8}$$
$$R=16.25\approx16.3$$

14. E

$$\frac{2x.5x}{2}=20$$
$$5x^2=20$$
$$x^2=4\Rightarrow x=2$$
$$h^2=4^2+10^2=\sqrt{116}=2\sqrt{29}$$

15. D

$$A+C+D=B+C+A\Rightarrow A=B=1\ ft^2=144in^2$$

16. C

$$x^2=16^2-8^2\Rightarrow x=8\sqrt{3}$$
$$x=8\sqrt{3}\ in=|1|Sin.$$

17. C

$$A=\frac{r}{p}=\frac{8\times10^8}{4\times10^{-4}}=2\times10^{12}$$

18. E

$$m=\frac{-\sqrt{5}}{5}=-0.45$$

19. D

$$|x-4|\geq8\Rightarrow x-4\geq8\ or\ x-4\leq-8\ so\ x\geq12\ or\ x\leq-4$$
-5, 12, 15 ⇒ Three minutes.

20. B

when $f(x)$ is a max, $x=\frac{\sqrt{3}}{12}=-0.866$

21. B

$$|AB|>|AC|\Rightarrow \hat{C}>\hat{B}\Rightarrow135°^2-a>45°\Rightarrow a<90°$$
Since a is positive, $0<a<90°$

22. E

The maximum member of intersection points is
$$3.4+\binom{3}{2}=12+3=15$$

23. C

Largest of such rectangles is the square.
Since a=b, $x.\sin65°=(20-x).\sin25°$
$$\Rightarrow x=\frac{20\sin25°}{\sin25°+\sin65°}=6.36\Rightarrow a=6.36\times\sin65=5.76$$
$\Rightarrow$ one side of the square is $11.53\Rightarrow$ Area = 133

24. A

$$c-2=0\Rightarrow c=2$$
$$\Rightarrow f(x)=3\Rightarrow f(2007)=3$$

25. E

$$y=x^2+x-3$$
$$3x=x^2+x-3$$
$$x^2-2x-3=0$$
$$(x-3)(x+1)=0\Rightarrow x=3\ or\ x=-1$$

26. D

Greatest area of DACB is $5.\frac{5}{2}.\frac{1}{2}=1.25$

27. E

The triangle is isosceles and right.

28. B

slope is $\dfrac{-1}{\dfrac{3--5}{-1-2}}=\dfrac{3}{8}=0.375$

29. A

The graph corresponds to $y=|x-2|$

30. A

$$2.0-0+6-2k=0$$
$$2k=6\Rightarrow k=3$$

31. A

$$\frac{1}{4}(-64)^{\frac{2}{3}}+(-64)^{\frac{1}{3}}=\frac{1}{4}.16-4=4-4=0$$

32. C

$$3^{2x-6}=3^{-2x}\Rightarrow2x-6=-2x\ so\ 4x=6\ and\ x=\frac{6}{4}=\frac{3}{2}$$

33. A

$x^2+y^2=25$ represents a circle; center of $(0,0)$ and $R=5$

34. B

$$\frac{\sqrt{18}+\sqrt{12}}{\sqrt{6}}=\frac{\sqrt{18}}{\sqrt{6}}+\frac{\sqrt{12}}{\sqrt{6}}=\sqrt{\frac{18}{6}}+\sqrt{\frac{18}{6}}=\sqrt{3}+\sqrt{2}$$

35. C

slope $=\dfrac{4-0}{3-1}=\dfrac{4}{2}=2$

$\Rightarrow y=2x+b \Rightarrow 0=2+b \Rightarrow b=-2$

$\Rightarrow y=2x-2 \Rightarrow y-2x+2=0$

36. B

$\Delta < 0-$

$(-A)^2-4.1.A < 0$

$A^2-4a < 0 \Rightarrow 0 < A < 4$

37. D

$f(x)=ax+b$

$f(x-1)+f(x+1)=6x-10$

$a(x-1)+b4a(x+19+b=6x-10$

$2ax+2b=6x-10 \Rightarrow a=\dfrac{-10}{2}=-5 \Rightarrow f(x)=3x-5$

$f(1)=3-5=-2$

38. D

$x= 0.2161616....$

$10x = 2.1616161...$

$100x=216.161616...$

$1000x-10x = 216-2 = 214$

$990x = 214$

$x =\dfrac{214}{990}$

39. C

Minor arc AB measures $180°-80°=100°$

Major arc Ab measures $360°-100°=260°$

40. E

$12^3 = 22.10.h \Rightarrow h =\dfrac{12^3}{220}= 7.85$

41. A

$V_{ar} =\dfrac{t+t}{\dfrac{t}{60}+\dfrac{t}{80}}=\dfrac{2.60.80}{140}= 68.57$

42. B

product of the roots is $\dfrac{-\sqrt{2}}{\sqrt{11}}= -0.426$

43. B

I and III imply that x is negative

II only implies that x is positive.

44. E

Each exterior angle is $180°-160°=20°$

$\Rightarrow$ Number of sides $=\dfrac{360°}{20°} = 18 \Rightarrow\dfrac{18.15}{2}= 135$

diagonals.

45. E

$(x-5)^2 =\left(4\sqrt{x}\right)^2$

$x^2-10x+25=16x$

$x^2-26x+25=0$

$(x-25)(x-1)=0 \Rightarrow$ x=25 and x=1. Only x=25 satisfies the original equation.

46. D

The graph intersect at 3 points.

47. E

Bull's total fazing area is.

$\pi.20^3.\dfrac{3}{4}+\pi.10^2.\dfrac{1}{4}= 1021 \approx 1020$

48. E

When $x \leq 1$, $\dfrac{x}{1}+\dfrac{y}{2}= 1$

$\Rightarrow\dfrac{y}{2}= 1-x$

$y=-2x+2 \Rightarrow g(x)=-2x+2$

$g^{-1}(x) =\dfrac{x-2}{-2}=\dfrac{2-x}{2}$

49. C

$\left(\dfrac{\dfrac{1}{h}+\dfrac{1}{24}}{3}+\dfrac{1}{\dfrac{24}{5}}\right).t=3 \Rightarrow\dfrac{11}{2h}.t=3 \Rightarrow t=\dfrac{64}{11}$

50. D

Squares with sides of length 1 → 25 of them

Squares with sides of length 2 → 16 of them

Squares with sides of length 3 → 9 of them

Squares with sides of length 4 → 4 of them

Squares with sides of length 5 → 1 of them

So there are totally 55 squares in the figure given.

Model Test 1 Answer Sheet

1.	Ⓐ	Ⓑ	Ⓒ	Ⓓ	Ⓔ		26.	Ⓐ	Ⓑ	Ⓒ	Ⓓ	Ⓔ
2.	Ⓐ	Ⓑ	Ⓒ	Ⓓ	Ⓔ		27.	Ⓐ	Ⓑ	Ⓒ	Ⓓ	Ⓔ
3.	Ⓐ	Ⓑ	Ⓒ	Ⓓ	Ⓔ		28.	Ⓐ	Ⓑ	Ⓒ	Ⓓ	Ⓔ
4.	Ⓐ	Ⓑ	Ⓒ	Ⓓ	Ⓔ		29.	Ⓐ	Ⓑ	Ⓒ	Ⓓ	Ⓔ
5.	Ⓐ	Ⓑ	Ⓒ	Ⓓ	Ⓔ		30.	Ⓐ	Ⓑ	Ⓒ	Ⓓ	Ⓔ
6.	Ⓐ	Ⓑ	Ⓒ	Ⓓ	Ⓔ		31.	Ⓐ	Ⓑ	Ⓒ	Ⓓ	Ⓔ
7.	Ⓐ	Ⓑ	Ⓒ	Ⓓ	Ⓔ		32.	Ⓐ	Ⓑ	Ⓒ	Ⓓ	Ⓔ
8.	Ⓐ	Ⓑ	Ⓒ	Ⓓ	Ⓔ		33.	Ⓐ	Ⓑ	Ⓒ	Ⓓ	Ⓔ
9.	Ⓐ	Ⓑ	Ⓒ	Ⓓ	Ⓔ		34.	Ⓐ	Ⓑ	Ⓒ	Ⓓ	Ⓔ
10.	Ⓐ	Ⓑ	Ⓒ	Ⓓ	Ⓔ		35.	Ⓐ	Ⓑ	Ⓒ	Ⓓ	Ⓔ
11.	Ⓐ	Ⓑ	Ⓒ	Ⓓ	Ⓔ		36.	Ⓐ	Ⓑ	Ⓒ	Ⓓ	Ⓔ
12.	Ⓐ	Ⓑ	Ⓒ	Ⓓ	Ⓔ		37.	Ⓐ	Ⓑ	Ⓒ	Ⓓ	Ⓔ
13.	Ⓐ	Ⓑ	Ⓒ	Ⓓ	Ⓔ		38.	Ⓐ	Ⓑ	Ⓒ	Ⓓ	Ⓔ
14.	Ⓐ	Ⓑ	Ⓒ	Ⓓ	Ⓔ		39.	Ⓐ	Ⓑ	Ⓒ	Ⓓ	Ⓔ
15.	Ⓐ	Ⓑ	Ⓒ	Ⓓ	Ⓔ		40.	Ⓐ	Ⓑ	Ⓒ	Ⓓ	Ⓔ
16.	Ⓐ	Ⓑ	Ⓒ	Ⓓ	Ⓔ		41.	Ⓐ	Ⓑ	Ⓒ	Ⓓ	Ⓔ
17.	Ⓐ	Ⓑ	Ⓒ	Ⓓ	Ⓔ		42.	Ⓐ	Ⓑ	Ⓒ	Ⓓ	Ⓔ
18.	Ⓐ	Ⓑ	Ⓒ	Ⓓ	Ⓔ		43.	Ⓐ	Ⓑ	Ⓒ	Ⓓ	Ⓔ
19.	Ⓐ	Ⓑ	Ⓒ	Ⓓ	Ⓔ		44.	Ⓐ	Ⓑ	Ⓒ	Ⓓ	Ⓔ
20.	Ⓐ	Ⓑ	Ⓒ	Ⓓ	Ⓔ		45.	Ⓐ	Ⓑ	Ⓒ	Ⓓ	Ⓔ
21.	Ⓐ	Ⓑ	Ⓒ	Ⓓ	Ⓔ		46.	Ⓐ	Ⓑ	Ⓒ	Ⓓ	Ⓔ
22.	Ⓐ	Ⓑ	Ⓒ	Ⓓ	Ⓔ		47.	Ⓐ	Ⓑ	Ⓒ	Ⓓ	Ⓔ
23.	Ⓐ	Ⓑ	Ⓒ	Ⓓ	Ⓔ		48.	Ⓐ	Ⓑ	Ⓒ	Ⓓ	Ⓔ
24.	Ⓐ	Ⓑ	Ⓒ	Ⓓ	Ⓔ		49.	Ⓐ	Ⓑ	Ⓒ	Ⓓ	Ⓔ
25.	Ⓐ	Ⓑ	Ⓒ	Ⓓ	Ⓔ		50.	Ⓐ	Ⓑ	Ⓒ	Ⓓ	Ⓔ

Model Test 2 Answer Sheet

1.	Ⓐ	Ⓑ	Ⓒ	Ⓓ	Ⓔ		26.	Ⓐ	Ⓑ	Ⓒ	Ⓓ	Ⓔ
2.	Ⓐ	Ⓑ	Ⓒ	Ⓓ	Ⓔ		27.	Ⓐ	Ⓑ	Ⓒ	Ⓓ	Ⓔ
3.	Ⓐ	Ⓑ	Ⓒ	Ⓓ	Ⓔ		28.	Ⓐ	Ⓑ	Ⓒ	Ⓓ	Ⓔ
4.	Ⓐ	Ⓑ	Ⓒ	Ⓓ	Ⓔ		29.	Ⓐ	Ⓑ	Ⓒ	Ⓓ	Ⓔ
5.	Ⓐ	Ⓑ	Ⓒ	Ⓓ	Ⓔ		30.	Ⓐ	Ⓑ	Ⓒ	Ⓓ	Ⓔ
6.	Ⓐ	Ⓑ	Ⓒ	Ⓓ	Ⓔ		31.	Ⓐ	Ⓑ	Ⓒ	Ⓓ	Ⓔ
7.	Ⓐ	Ⓑ	Ⓒ	Ⓓ	Ⓔ		32.	Ⓐ	Ⓑ	Ⓒ	Ⓓ	Ⓔ
8.	Ⓐ	Ⓑ	Ⓒ	Ⓓ	Ⓔ		33.	Ⓐ	Ⓑ	Ⓒ	Ⓓ	Ⓔ
9.	Ⓐ	Ⓑ	Ⓒ	Ⓓ	Ⓔ		34.	Ⓐ	Ⓑ	Ⓒ	Ⓓ	Ⓔ
10.	Ⓐ	Ⓑ	Ⓒ	Ⓓ	Ⓔ		35.	Ⓐ	Ⓑ	Ⓒ	Ⓓ	Ⓔ
11.	Ⓐ	Ⓑ	Ⓒ	Ⓓ	Ⓔ		36.	Ⓐ	Ⓑ	Ⓒ	Ⓓ	Ⓔ
12.	Ⓐ	Ⓑ	Ⓒ	Ⓓ	Ⓔ		37.	Ⓐ	Ⓑ	Ⓒ	Ⓓ	Ⓔ
13.	Ⓐ	Ⓑ	Ⓒ	Ⓓ	Ⓔ		38.	Ⓐ	Ⓑ	Ⓒ	Ⓓ	Ⓔ
14.	Ⓐ	Ⓑ	Ⓒ	Ⓓ	Ⓔ		39.	Ⓐ	Ⓑ	Ⓒ	Ⓓ	Ⓔ
15.	Ⓐ	Ⓑ	Ⓒ	Ⓓ	Ⓔ		40.	Ⓐ	Ⓑ	Ⓒ	Ⓓ	Ⓔ
16.	Ⓐ	Ⓑ	Ⓒ	Ⓓ	Ⓔ		41.	Ⓐ	Ⓑ	Ⓒ	Ⓓ	Ⓔ
17.	Ⓐ	Ⓑ	Ⓒ	Ⓓ	Ⓔ		42.	Ⓐ	Ⓑ	Ⓒ	Ⓓ	Ⓔ
18.	Ⓐ	Ⓑ	Ⓒ	Ⓓ	Ⓔ		43.	Ⓐ	Ⓑ	Ⓒ	Ⓓ	Ⓔ
19.	Ⓐ	Ⓑ	Ⓒ	Ⓓ	Ⓔ		44.	Ⓐ	Ⓑ	Ⓒ	Ⓓ	Ⓔ
20.	Ⓐ	Ⓑ	Ⓒ	Ⓓ	Ⓔ		45.	Ⓐ	Ⓑ	Ⓒ	Ⓓ	Ⓔ
21.	Ⓐ	Ⓑ	Ⓒ	Ⓓ	Ⓔ		46.	Ⓐ	Ⓑ	Ⓒ	Ⓓ	Ⓔ
22.	Ⓐ	Ⓑ	Ⓒ	Ⓓ	Ⓔ		47.	Ⓐ	Ⓑ	Ⓒ	Ⓓ	Ⓔ
23.	Ⓐ	Ⓑ	Ⓒ	Ⓓ	Ⓔ		48.	Ⓐ	Ⓑ	Ⓒ	Ⓓ	Ⓔ
24.	Ⓐ	Ⓑ	Ⓒ	Ⓓ	Ⓔ		49.	Ⓐ	Ⓑ	Ⓒ	Ⓓ	Ⓔ
25.	Ⓐ	Ⓑ	Ⓒ	Ⓓ	Ⓔ		50.	Ⓐ	Ⓑ	Ⓒ	Ⓓ	Ⓔ

Model Test 3 Answer Sheet

#						#					
1.	Ⓐ	Ⓑ	Ⓒ	Ⓓ	Ⓔ	26.	Ⓐ	Ⓑ	Ⓒ	Ⓓ	Ⓔ
2.	Ⓐ	Ⓑ	Ⓒ	Ⓓ	Ⓔ	27.	Ⓐ	Ⓑ	Ⓒ	Ⓓ	Ⓔ
3.	Ⓐ	Ⓑ	Ⓒ	Ⓓ	Ⓔ	28.	Ⓐ	Ⓑ	Ⓒ	Ⓓ	Ⓔ
4.	Ⓐ	Ⓑ	Ⓒ	Ⓓ	Ⓔ	29.	Ⓐ	Ⓑ	Ⓒ	Ⓓ	Ⓔ
5.	Ⓐ	Ⓑ	Ⓒ	Ⓓ	Ⓔ	30.	Ⓐ	Ⓑ	Ⓒ	Ⓓ	Ⓔ
6.	Ⓐ	Ⓑ	Ⓒ	Ⓓ	Ⓔ	31.	Ⓐ	Ⓑ	Ⓒ	Ⓓ	Ⓔ
7.	Ⓐ	Ⓑ	Ⓒ	Ⓓ	Ⓔ	32.	Ⓐ	Ⓑ	Ⓒ	Ⓓ	Ⓔ
8.	Ⓐ	Ⓑ	Ⓒ	Ⓓ	Ⓔ	33.	Ⓐ	Ⓑ	Ⓒ	Ⓓ	Ⓔ
9.	Ⓐ	Ⓑ	Ⓒ	Ⓓ	Ⓔ	34.	Ⓐ	Ⓑ	Ⓒ	Ⓓ	Ⓔ
10.	Ⓐ	Ⓑ	Ⓒ	Ⓓ	Ⓔ	35.	Ⓐ	Ⓑ	Ⓒ	Ⓓ	Ⓔ
11.	Ⓐ	Ⓑ	Ⓒ	Ⓓ	Ⓔ	36.	Ⓐ	Ⓑ	Ⓒ	Ⓓ	Ⓔ
12.	Ⓐ	Ⓑ	Ⓒ	Ⓓ	Ⓔ	37.	Ⓐ	Ⓑ	Ⓒ	Ⓓ	Ⓔ
13.	Ⓐ	Ⓑ	Ⓒ	Ⓓ	Ⓔ	38.	Ⓐ	Ⓑ	Ⓒ	Ⓓ	Ⓔ
14.	Ⓐ	Ⓑ	Ⓒ	Ⓓ	Ⓔ	39.	Ⓐ	Ⓑ	Ⓒ	Ⓓ	Ⓔ
15.	Ⓐ	Ⓑ	Ⓒ	Ⓓ	Ⓔ	40.	Ⓐ	Ⓑ	Ⓒ	Ⓓ	Ⓔ
16.	Ⓐ	Ⓑ	Ⓒ	Ⓓ	Ⓔ	41.	Ⓐ	Ⓑ	Ⓒ	Ⓓ	Ⓔ
17.	Ⓐ	Ⓑ	Ⓒ	Ⓓ	Ⓔ	42.	Ⓐ	Ⓑ	Ⓒ	Ⓓ	Ⓔ
18.	Ⓐ	Ⓑ	Ⓒ	Ⓓ	Ⓔ	43.	Ⓐ	Ⓑ	Ⓒ	Ⓓ	Ⓔ
19.	Ⓐ	Ⓑ	Ⓒ	Ⓓ	Ⓔ	44.	Ⓐ	Ⓑ	Ⓒ	Ⓓ	Ⓔ
20.	Ⓐ	Ⓑ	Ⓒ	Ⓓ	Ⓔ	45.	Ⓐ	Ⓑ	Ⓒ	Ⓓ	Ⓔ
21.	Ⓐ	Ⓑ	Ⓒ	Ⓓ	Ⓔ	46.	Ⓐ	Ⓑ	Ⓒ	Ⓓ	Ⓔ
22.	Ⓐ	Ⓑ	Ⓒ	Ⓓ	Ⓔ	47.	Ⓐ	Ⓑ	Ⓒ	Ⓓ	Ⓔ
23.	Ⓐ	Ⓑ	Ⓒ	Ⓓ	Ⓔ	48.	Ⓐ	Ⓑ	Ⓒ	Ⓓ	Ⓔ
24.	Ⓐ	Ⓑ	Ⓒ	Ⓓ	Ⓔ	49.	Ⓐ	Ⓑ	Ⓒ	Ⓓ	Ⓔ
25.	Ⓐ	Ⓑ	Ⓒ	Ⓓ	Ⓔ	50.	Ⓐ	Ⓑ	Ⓒ	Ⓓ	Ⓔ

Model Test 4 Answer Sheet

#						#					
1.	Ⓐ	Ⓑ	Ⓒ	Ⓓ	Ⓔ	26.	Ⓐ	Ⓑ	Ⓒ	Ⓓ	Ⓔ
2.	Ⓐ	Ⓑ	Ⓒ	Ⓓ	Ⓔ	27.	Ⓐ	Ⓑ	Ⓒ	Ⓓ	Ⓔ
3.	Ⓐ	Ⓑ	Ⓒ	Ⓓ	Ⓔ	28.	Ⓐ	Ⓑ	Ⓒ	Ⓓ	Ⓔ
4.	Ⓐ	Ⓑ	Ⓒ	Ⓓ	Ⓔ	29.	Ⓐ	Ⓑ	Ⓒ	Ⓓ	Ⓔ
5.	Ⓐ	Ⓑ	Ⓒ	Ⓓ	Ⓔ	30.	Ⓐ	Ⓑ	Ⓒ	Ⓓ	Ⓔ
6.	Ⓐ	Ⓑ	Ⓒ	Ⓓ	Ⓔ	31.	Ⓐ	Ⓑ	Ⓒ	Ⓓ	Ⓔ
7.	Ⓐ	Ⓑ	Ⓒ	Ⓓ	Ⓔ	32.	Ⓐ	Ⓑ	Ⓒ	Ⓓ	Ⓔ
8.	Ⓐ	Ⓑ	Ⓒ	Ⓓ	Ⓔ	33.	Ⓐ	Ⓑ	Ⓒ	Ⓓ	Ⓔ
9.	Ⓐ	Ⓑ	Ⓒ	Ⓓ	Ⓔ	34.	Ⓐ	Ⓑ	Ⓒ	Ⓓ	Ⓔ
10.	Ⓐ	Ⓑ	Ⓒ	Ⓓ	Ⓔ	35.	Ⓐ	Ⓑ	Ⓒ	Ⓓ	Ⓔ
11.	Ⓐ	Ⓑ	Ⓒ	Ⓓ	Ⓔ	36.	Ⓐ	Ⓑ	Ⓒ	Ⓓ	Ⓔ
12.	Ⓐ	Ⓑ	Ⓒ	Ⓓ	Ⓔ	37.	Ⓐ	Ⓑ	Ⓒ	Ⓓ	Ⓔ
13.	Ⓐ	Ⓑ	Ⓒ	Ⓓ	Ⓔ	38.	Ⓐ	Ⓑ	Ⓒ	Ⓓ	Ⓔ
14.	Ⓐ	Ⓑ	Ⓒ	Ⓓ	Ⓔ	39.	Ⓐ	Ⓑ	Ⓒ	Ⓓ	Ⓔ
15.	Ⓐ	Ⓑ	Ⓒ	Ⓓ	Ⓔ	40.	Ⓐ	Ⓑ	Ⓒ	Ⓓ	Ⓔ
16.	Ⓐ	Ⓑ	Ⓒ	Ⓓ	Ⓔ	41.	Ⓐ	Ⓑ	Ⓒ	Ⓓ	Ⓔ
17.	Ⓐ	Ⓑ	Ⓒ	Ⓓ	Ⓔ	42.	Ⓐ	Ⓑ	Ⓒ	Ⓓ	Ⓔ
18.	Ⓐ	Ⓑ	Ⓒ	Ⓓ	Ⓔ	43.	Ⓐ	Ⓑ	Ⓒ	Ⓓ	Ⓔ
19.	Ⓐ	Ⓑ	Ⓒ	Ⓓ	Ⓔ	44.	Ⓐ	Ⓑ	Ⓒ	Ⓓ	Ⓔ
20.	Ⓐ	Ⓑ	Ⓒ	Ⓓ	Ⓔ	45.	Ⓐ	Ⓑ	Ⓒ	Ⓓ	Ⓔ
21.	Ⓐ	Ⓑ	Ⓒ	Ⓓ	Ⓔ	46.	Ⓐ	Ⓑ	Ⓒ	Ⓓ	Ⓔ
22.	Ⓐ	Ⓑ	Ⓒ	Ⓓ	Ⓔ	47.	Ⓐ	Ⓑ	Ⓒ	Ⓓ	Ⓔ
23.	Ⓐ	Ⓑ	Ⓒ	Ⓓ	Ⓔ	48.	Ⓐ	Ⓑ	Ⓒ	Ⓓ	Ⓔ
24.	Ⓐ	Ⓑ	Ⓒ	Ⓓ	Ⓔ	49.	Ⓐ	Ⓑ	Ⓒ	Ⓓ	Ⓔ
25.	Ⓐ	Ⓑ	Ⓒ	Ⓓ	Ⓔ	50.	Ⓐ	Ⓑ	Ⓒ	Ⓓ	Ⓔ

Model Test 5 Answer Sheet

1.	Ⓐ Ⓑ Ⓒ Ⓓ Ⓔ	26. Ⓐ Ⓑ Ⓒ Ⓓ Ⓔ
2.	Ⓐ Ⓑ Ⓒ Ⓓ Ⓔ	27. Ⓐ Ⓑ Ⓒ Ⓓ Ⓔ
3.	Ⓐ Ⓑ Ⓒ Ⓓ Ⓔ	28. Ⓐ Ⓑ Ⓒ Ⓓ Ⓔ
4.	Ⓐ Ⓑ Ⓒ Ⓓ Ⓔ	29. Ⓐ Ⓑ Ⓒ Ⓓ Ⓔ
5.	Ⓐ Ⓑ Ⓒ Ⓓ Ⓔ	30. Ⓐ Ⓑ Ⓒ Ⓓ Ⓔ
6.	Ⓐ Ⓑ Ⓒ Ⓓ Ⓔ	31. Ⓐ Ⓑ Ⓒ Ⓓ Ⓔ
7.	Ⓐ Ⓑ Ⓒ Ⓓ Ⓔ	32. Ⓐ Ⓑ Ⓒ Ⓓ Ⓔ
8.	Ⓐ Ⓑ Ⓒ Ⓓ Ⓔ	33. Ⓐ Ⓑ Ⓒ Ⓓ Ⓔ
9.	Ⓐ Ⓑ Ⓒ Ⓓ Ⓔ	34. Ⓐ Ⓑ Ⓒ Ⓓ Ⓔ
10.	Ⓐ Ⓑ Ⓒ Ⓓ Ⓔ	35. Ⓐ Ⓑ Ⓒ Ⓓ Ⓔ
11.	Ⓐ Ⓑ Ⓒ Ⓓ Ⓔ	36. Ⓐ Ⓑ Ⓒ Ⓓ Ⓔ
12.	Ⓐ Ⓑ Ⓒ Ⓓ Ⓔ	37. Ⓐ Ⓑ Ⓒ Ⓓ Ⓔ
13.	Ⓐ Ⓑ Ⓒ Ⓓ Ⓔ	38. Ⓐ Ⓑ Ⓒ Ⓓ Ⓔ
14.	Ⓐ Ⓑ Ⓒ Ⓓ Ⓔ	39. Ⓐ Ⓑ Ⓒ Ⓓ Ⓔ
15.	Ⓐ Ⓑ Ⓒ Ⓓ Ⓔ	40. Ⓐ Ⓑ Ⓒ Ⓓ Ⓔ
16.	Ⓐ Ⓑ Ⓒ Ⓓ Ⓔ	41. Ⓐ Ⓑ Ⓒ Ⓓ Ⓔ
17.	Ⓐ Ⓑ Ⓒ Ⓓ Ⓔ	42. Ⓐ Ⓑ Ⓒ Ⓓ Ⓔ
18.	Ⓐ Ⓑ Ⓒ Ⓓ Ⓔ	43. Ⓐ Ⓑ Ⓒ Ⓓ Ⓔ
19.	Ⓐ Ⓑ Ⓒ Ⓓ Ⓔ	44. Ⓐ Ⓑ Ⓒ Ⓓ Ⓔ
20.	Ⓐ Ⓑ Ⓒ Ⓓ Ⓔ	45. Ⓐ Ⓑ Ⓒ Ⓓ Ⓔ
21.	Ⓐ Ⓑ Ⓒ Ⓓ Ⓔ	46. Ⓐ Ⓑ Ⓒ Ⓓ Ⓔ
22.	Ⓐ Ⓑ Ⓒ Ⓓ Ⓔ	47. Ⓐ Ⓑ Ⓒ Ⓓ Ⓔ
23.	Ⓐ Ⓑ Ⓒ Ⓓ Ⓔ	48. Ⓐ Ⓑ Ⓒ Ⓓ Ⓔ
24.	Ⓐ Ⓑ Ⓒ Ⓓ Ⓔ	49. Ⓐ Ⓑ Ⓒ Ⓓ Ⓔ
25.	Ⓐ Ⓑ Ⓒ Ⓓ Ⓔ	50. Ⓐ Ⓑ Ⓒ Ⓓ Ⓔ

Model Test 6 Answer Sheet

1.	Ⓐ Ⓑ Ⓒ Ⓓ Ⓔ	26. Ⓐ Ⓑ Ⓒ Ⓓ Ⓔ
2.	Ⓐ Ⓑ Ⓒ Ⓓ Ⓔ	27. Ⓐ Ⓑ Ⓒ Ⓓ Ⓔ
3.	Ⓐ Ⓑ Ⓒ Ⓓ Ⓔ	28. Ⓐ Ⓑ Ⓒ Ⓓ Ⓔ
4.	Ⓐ Ⓑ Ⓒ Ⓓ Ⓔ	29. Ⓐ Ⓑ Ⓒ Ⓓ Ⓔ
5.	Ⓐ Ⓑ Ⓒ Ⓓ Ⓔ	30. Ⓐ Ⓑ Ⓒ Ⓓ Ⓔ
6.	Ⓐ Ⓑ Ⓒ Ⓓ Ⓔ	31. Ⓐ Ⓑ Ⓒ Ⓓ Ⓔ
7.	Ⓐ Ⓑ Ⓒ Ⓓ Ⓔ	32. Ⓐ Ⓑ Ⓒ Ⓓ Ⓔ
8.	Ⓐ Ⓑ Ⓒ Ⓓ Ⓔ	33. Ⓐ Ⓑ Ⓒ Ⓓ Ⓔ
9.	Ⓐ Ⓑ Ⓒ Ⓓ Ⓔ	34. Ⓐ Ⓑ Ⓒ Ⓓ Ⓔ
10.	Ⓐ Ⓑ Ⓒ Ⓓ Ⓔ	35. Ⓐ Ⓑ Ⓒ Ⓓ Ⓔ
11.	Ⓐ Ⓑ Ⓒ Ⓓ Ⓔ	36. Ⓐ Ⓑ Ⓒ Ⓓ Ⓔ
12.	Ⓐ Ⓑ Ⓒ Ⓓ Ⓔ	37. Ⓐ Ⓑ Ⓒ Ⓓ Ⓔ
13.	Ⓐ Ⓑ Ⓒ Ⓓ Ⓔ	38. Ⓐ Ⓑ Ⓒ Ⓓ Ⓔ
14.	Ⓐ Ⓑ Ⓒ Ⓓ Ⓔ	39. Ⓐ Ⓑ Ⓒ Ⓓ Ⓔ
15.	Ⓐ Ⓑ Ⓒ Ⓓ Ⓔ	40. Ⓐ Ⓑ Ⓒ Ⓓ Ⓔ
16.	Ⓐ Ⓑ Ⓒ Ⓓ Ⓔ	41. Ⓐ Ⓑ Ⓒ Ⓓ Ⓔ
17.	Ⓐ Ⓑ Ⓒ Ⓓ Ⓔ	42. Ⓐ Ⓑ Ⓒ Ⓓ Ⓔ
18.	Ⓐ Ⓑ Ⓒ Ⓓ Ⓔ	43. Ⓐ Ⓑ Ⓒ Ⓓ Ⓔ
19.	Ⓐ Ⓑ Ⓒ Ⓓ Ⓔ	44. Ⓐ Ⓑ Ⓒ Ⓓ Ⓔ
20.	Ⓐ Ⓑ Ⓒ Ⓓ Ⓔ	45. Ⓐ Ⓑ Ⓒ Ⓓ Ⓔ
21.	Ⓐ Ⓑ Ⓒ Ⓓ Ⓔ	46. Ⓐ Ⓑ Ⓒ Ⓓ Ⓔ
22.	Ⓐ Ⓑ Ⓒ Ⓓ Ⓔ	47. Ⓐ Ⓑ Ⓒ Ⓓ Ⓔ
23.	Ⓐ Ⓑ Ⓒ Ⓓ Ⓔ	48. Ⓐ Ⓑ Ⓒ Ⓓ Ⓔ
24.	Ⓐ Ⓑ Ⓒ Ⓓ Ⓔ	49. Ⓐ Ⓑ Ⓒ Ⓓ Ⓔ
25.	Ⓐ Ⓑ Ⓒ Ⓓ Ⓔ	50. Ⓐ Ⓑ Ⓒ Ⓓ Ⓔ

Model Test 7 Answer Sheet

#							#					
1.	Ⓐ	Ⓑ	Ⓒ	Ⓓ	Ⓔ		26.	Ⓐ	Ⓑ	Ⓒ	Ⓓ	Ⓔ
2.	Ⓐ	Ⓑ	Ⓒ	Ⓓ	Ⓔ		27.	Ⓐ	Ⓑ	Ⓒ	Ⓓ	Ⓔ
3.	Ⓐ	Ⓑ	Ⓒ	Ⓓ	Ⓔ		28.	Ⓐ	Ⓑ	Ⓒ	Ⓓ	Ⓔ
4.	Ⓐ	Ⓑ	Ⓒ	Ⓓ	Ⓔ		29.	Ⓐ	Ⓑ	Ⓒ	Ⓓ	Ⓔ
5.	Ⓐ	Ⓑ	Ⓒ	Ⓓ	Ⓔ		30.	Ⓐ	Ⓑ	Ⓒ	Ⓓ	Ⓔ
6.	Ⓐ	Ⓑ	Ⓒ	Ⓓ	Ⓔ		31.	Ⓐ	Ⓑ	Ⓒ	Ⓓ	Ⓔ
7.	Ⓐ	Ⓑ	Ⓒ	Ⓓ	Ⓔ		32.	Ⓐ	Ⓑ	Ⓒ	Ⓓ	Ⓔ
8.	Ⓐ	Ⓑ	Ⓒ	Ⓓ	Ⓔ		33.	Ⓐ	Ⓑ	Ⓒ	Ⓓ	Ⓔ
9.	Ⓐ	Ⓑ	Ⓒ	Ⓓ	Ⓔ		34.	Ⓐ	Ⓑ	Ⓒ	Ⓓ	Ⓔ
10.	Ⓐ	Ⓑ	Ⓒ	Ⓓ	Ⓔ		35.	Ⓐ	Ⓑ	Ⓒ	Ⓓ	Ⓔ
11.	Ⓐ	Ⓑ	Ⓒ	Ⓓ	Ⓔ		36.	Ⓐ	Ⓑ	Ⓒ	Ⓓ	Ⓔ
12.	Ⓐ	Ⓑ	Ⓒ	Ⓓ	Ⓔ		37.	Ⓐ	Ⓑ	Ⓒ	Ⓓ	Ⓔ
13.	Ⓐ	Ⓑ	Ⓒ	Ⓓ	Ⓔ		38.	Ⓐ	Ⓑ	Ⓒ	Ⓓ	Ⓔ
14.	Ⓐ	Ⓑ	Ⓒ	Ⓓ	Ⓔ		39.	Ⓐ	Ⓑ	Ⓒ	Ⓓ	Ⓔ
15.	Ⓐ	Ⓑ	Ⓒ	Ⓓ	Ⓔ		40.	Ⓐ	Ⓑ	Ⓒ	Ⓓ	Ⓔ
16.	Ⓐ	Ⓑ	Ⓒ	Ⓓ	Ⓔ		41.	Ⓐ	Ⓑ	Ⓒ	Ⓓ	Ⓔ
17.	Ⓐ	Ⓑ	Ⓒ	Ⓓ	Ⓔ		42.	Ⓐ	Ⓑ	Ⓒ	Ⓓ	Ⓔ
18.	Ⓐ	Ⓑ	Ⓒ	Ⓓ	Ⓔ		43.	Ⓐ	Ⓑ	Ⓒ	Ⓓ	Ⓔ
19.	Ⓐ	Ⓑ	Ⓒ	Ⓓ	Ⓔ		44.	Ⓐ	Ⓑ	Ⓒ	Ⓓ	Ⓔ
20.	Ⓐ	Ⓑ	Ⓒ	Ⓓ	Ⓔ		45.	Ⓐ	Ⓑ	Ⓒ	Ⓓ	Ⓔ
21.	Ⓐ	Ⓑ	Ⓒ	Ⓓ	Ⓔ		46.	Ⓐ	Ⓑ	Ⓒ	Ⓓ	Ⓔ
22.	Ⓐ	Ⓑ	Ⓒ	Ⓓ	Ⓔ		47.	Ⓐ	Ⓑ	Ⓒ	Ⓓ	Ⓔ
23.	Ⓐ	Ⓑ	Ⓒ	Ⓓ	Ⓔ		48.	Ⓐ	Ⓑ	Ⓒ	Ⓓ	Ⓔ
24.	Ⓐ	Ⓑ	Ⓒ	Ⓓ	Ⓔ		49.	Ⓐ	Ⓑ	Ⓒ	Ⓓ	Ⓔ
25.	Ⓐ	Ⓑ	Ⓒ	Ⓓ	Ⓔ		50.	Ⓐ	Ⓑ	Ⓒ	Ⓓ	Ⓔ

Model Test 8 Answer Sheet

#							#					
1.	Ⓐ	Ⓑ	Ⓒ	Ⓓ	Ⓔ		26.	Ⓐ	Ⓑ	Ⓒ	Ⓓ	Ⓔ
2.	Ⓐ	Ⓑ	Ⓒ	Ⓓ	Ⓔ		27.	Ⓐ	Ⓑ	Ⓒ	Ⓓ	Ⓔ
3.	Ⓐ	Ⓑ	Ⓒ	Ⓓ	Ⓔ		28.	Ⓐ	Ⓑ	Ⓒ	Ⓓ	Ⓔ
4.	Ⓐ	Ⓑ	Ⓒ	Ⓓ	Ⓔ		29.	Ⓐ	Ⓑ	Ⓒ	Ⓓ	Ⓔ
5.	Ⓐ	Ⓑ	Ⓒ	Ⓓ	Ⓔ		30.	Ⓐ	Ⓑ	Ⓒ	Ⓓ	Ⓔ
6.	Ⓐ	Ⓑ	Ⓒ	Ⓓ	Ⓔ		31.	Ⓐ	Ⓑ	Ⓒ	Ⓓ	Ⓔ
7.	Ⓐ	Ⓑ	Ⓒ	Ⓓ	Ⓔ		32.	Ⓐ	Ⓑ	Ⓒ	Ⓓ	Ⓔ
8.	Ⓐ	Ⓑ	Ⓒ	Ⓓ	Ⓔ		33.	Ⓐ	Ⓑ	Ⓒ	Ⓓ	Ⓔ
9.	Ⓐ	Ⓑ	Ⓒ	Ⓓ	Ⓔ		34.	Ⓐ	Ⓑ	Ⓒ	Ⓓ	Ⓔ
10.	Ⓐ	Ⓑ	Ⓒ	Ⓓ	Ⓔ		35.	Ⓐ	Ⓑ	Ⓒ	Ⓓ	Ⓔ
11.	Ⓐ	Ⓑ	Ⓒ	Ⓓ	Ⓔ		36.	Ⓐ	Ⓑ	Ⓒ	Ⓓ	Ⓔ
12.	Ⓐ	Ⓑ	Ⓒ	Ⓓ	Ⓔ		37.	Ⓐ	Ⓑ	Ⓒ	Ⓓ	Ⓔ
13.	Ⓐ	Ⓑ	Ⓒ	Ⓓ	Ⓔ		38.	Ⓐ	Ⓑ	Ⓒ	Ⓓ	Ⓔ
14.	Ⓐ	Ⓑ	Ⓒ	Ⓓ	Ⓔ		39.	Ⓐ	Ⓑ	Ⓒ	Ⓓ	Ⓔ
15.	Ⓐ	Ⓑ	Ⓒ	Ⓓ	Ⓔ		40.	Ⓐ	Ⓑ	Ⓒ	Ⓓ	Ⓔ
16.	Ⓐ	Ⓑ	Ⓒ	Ⓓ	Ⓔ		41.	Ⓐ	Ⓑ	Ⓒ	Ⓓ	Ⓔ
17.	Ⓐ	Ⓑ	Ⓒ	Ⓓ	Ⓔ		42.	Ⓐ	Ⓑ	Ⓒ	Ⓓ	Ⓔ
18.	Ⓐ	Ⓑ	Ⓒ	Ⓓ	Ⓔ		43.	Ⓐ	Ⓑ	Ⓒ	Ⓓ	Ⓔ
19.	Ⓐ	Ⓑ	Ⓒ	Ⓓ	Ⓔ		44.	Ⓐ	Ⓑ	Ⓒ	Ⓓ	Ⓔ
20.	Ⓐ	Ⓑ	Ⓒ	Ⓓ	Ⓔ		45.	Ⓐ	Ⓑ	Ⓒ	Ⓓ	Ⓔ
21.	Ⓐ	Ⓑ	Ⓒ	Ⓓ	Ⓔ		46.	Ⓐ	Ⓑ	Ⓒ	Ⓓ	Ⓔ
22.	Ⓐ	Ⓑ	Ⓒ	Ⓓ	Ⓔ		47.	Ⓐ	Ⓑ	Ⓒ	Ⓓ	Ⓔ
23.	Ⓐ	Ⓑ	Ⓒ	Ⓓ	Ⓔ		48.	Ⓐ	Ⓑ	Ⓒ	Ⓓ	Ⓔ
24.	Ⓐ	Ⓑ	Ⓒ	Ⓓ	Ⓔ		49.	Ⓐ	Ⓑ	Ⓒ	Ⓓ	Ⓔ
25.	Ⓐ	Ⓑ	Ⓒ	Ⓓ	Ⓔ		50.	Ⓐ	Ⓑ	Ⓒ	Ⓓ	Ⓔ

Model Test 9 Answer Sheet

1.	Ⓐ	Ⓑ	Ⓒ	Ⓓ	Ⓔ
2.	Ⓐ	Ⓑ	Ⓒ	Ⓓ	Ⓔ
3.	Ⓐ	Ⓑ	Ⓒ	Ⓓ	Ⓔ
4.	Ⓐ	Ⓑ	Ⓒ	Ⓓ	Ⓔ
5.	Ⓐ	Ⓑ	Ⓒ	Ⓓ	Ⓔ
6.	Ⓐ	Ⓑ	Ⓒ	Ⓓ	Ⓔ
7.	Ⓐ	Ⓑ	Ⓒ	Ⓓ	Ⓔ
8.	Ⓐ	Ⓑ	Ⓒ	Ⓓ	Ⓔ
9.	Ⓐ	Ⓑ	Ⓒ	Ⓓ	Ⓔ
10.	Ⓐ	Ⓑ	Ⓒ	Ⓓ	Ⓔ
11.	Ⓐ	Ⓑ	Ⓒ	Ⓓ	Ⓔ
12.	Ⓐ	Ⓑ	Ⓒ	Ⓓ	Ⓔ
13.	Ⓐ	Ⓑ	Ⓒ	Ⓓ	Ⓔ
14.	Ⓐ	Ⓑ	Ⓒ	Ⓓ	Ⓔ
15.	Ⓐ	Ⓑ	Ⓒ	Ⓓ	Ⓔ
16.	Ⓐ	Ⓑ	Ⓒ	Ⓓ	Ⓔ
17.	Ⓐ	Ⓑ	Ⓒ	Ⓓ	Ⓔ
18.	Ⓐ	Ⓑ	Ⓒ	Ⓓ	Ⓔ
19.	Ⓐ	Ⓑ	Ⓒ	Ⓓ	Ⓔ
20.	Ⓐ	Ⓑ	Ⓒ	Ⓓ	Ⓔ
21.	Ⓐ	Ⓑ	Ⓒ	Ⓓ	Ⓔ
22.	Ⓐ	Ⓑ	Ⓒ	Ⓓ	Ⓔ
23.	Ⓐ	Ⓑ	Ⓒ	Ⓓ	Ⓔ
24.	Ⓐ	Ⓑ	Ⓒ	Ⓓ	Ⓔ
25.	Ⓐ	Ⓑ	Ⓒ	Ⓓ	Ⓔ
26.	Ⓐ	Ⓑ	Ⓒ	Ⓓ	Ⓔ
27.	Ⓐ	Ⓑ	Ⓒ	Ⓓ	Ⓔ
28.	Ⓐ	Ⓑ	Ⓒ	Ⓓ	Ⓔ
29.	Ⓐ	Ⓑ	Ⓒ	Ⓓ	Ⓔ
30.	Ⓐ	Ⓑ	Ⓒ	Ⓓ	Ⓔ
31.	Ⓐ	Ⓑ	Ⓒ	Ⓓ	Ⓔ
32.	Ⓐ	Ⓑ	Ⓒ	Ⓓ	Ⓔ
33.	Ⓐ	Ⓑ	Ⓒ	Ⓓ	Ⓔ
34.	Ⓐ	Ⓑ	Ⓒ	Ⓓ	Ⓔ
35.	Ⓐ	Ⓑ	Ⓒ	Ⓓ	Ⓔ
36.	Ⓐ	Ⓑ	Ⓒ	Ⓓ	Ⓔ
37.	Ⓐ	Ⓑ	Ⓒ	Ⓓ	Ⓔ
38.	Ⓐ	Ⓑ	Ⓒ	Ⓓ	Ⓔ
39.	Ⓐ	Ⓑ	Ⓒ	Ⓓ	Ⓔ
40.	Ⓐ	Ⓑ	Ⓒ	Ⓓ	Ⓔ
41.	Ⓐ	Ⓑ	Ⓒ	Ⓓ	Ⓔ
42.	Ⓐ	Ⓑ	Ⓒ	Ⓓ	Ⓔ
43.	Ⓐ	Ⓑ	Ⓒ	Ⓓ	Ⓔ
44.	Ⓐ	Ⓑ	Ⓒ	Ⓓ	Ⓔ
45.	Ⓐ	Ⓑ	Ⓒ	Ⓓ	Ⓔ
46.	Ⓐ	Ⓑ	Ⓒ	Ⓓ	Ⓔ
47.	Ⓐ	Ⓑ	Ⓒ	Ⓓ	Ⓔ
48.	Ⓐ	Ⓑ	Ⓒ	Ⓓ	Ⓔ
49.	Ⓐ	Ⓑ	Ⓒ	Ⓓ	Ⓔ
50.	Ⓐ	Ⓑ	Ⓒ	Ⓓ	Ⓔ

Model Test 10 Answer Sheet

1.	Ⓐ	Ⓑ	Ⓒ	Ⓓ	Ⓔ
2.	Ⓐ	Ⓑ	Ⓒ	Ⓓ	Ⓔ
3.	Ⓐ	Ⓑ	Ⓒ	Ⓓ	Ⓔ
4.	Ⓐ	Ⓑ	Ⓒ	Ⓓ	Ⓔ
5.	Ⓐ	Ⓑ	Ⓒ	Ⓓ	Ⓔ
6.	Ⓐ	Ⓑ	Ⓒ	Ⓓ	Ⓔ
7.	Ⓐ	Ⓑ	Ⓒ	Ⓓ	Ⓔ
8.	Ⓐ	Ⓑ	Ⓒ	Ⓓ	Ⓔ
9.	Ⓐ	Ⓑ	Ⓒ	Ⓓ	Ⓔ
10.	Ⓐ	Ⓑ	Ⓒ	Ⓓ	Ⓔ
11.	Ⓐ	Ⓑ	Ⓒ	Ⓓ	Ⓔ
12.	Ⓐ	Ⓑ	Ⓒ	Ⓓ	Ⓔ
13.	Ⓐ	Ⓑ	Ⓒ	Ⓓ	Ⓔ
14.	Ⓐ	Ⓑ	Ⓒ	Ⓓ	Ⓔ
15.	Ⓐ	Ⓑ	Ⓒ	Ⓓ	Ⓔ
16.	Ⓐ	Ⓑ	Ⓒ	Ⓓ	Ⓔ
17.	Ⓐ	Ⓑ	Ⓒ	Ⓓ	Ⓔ
18.	Ⓐ	Ⓑ	Ⓒ	Ⓓ	Ⓔ
19.	Ⓐ	Ⓑ	Ⓒ	Ⓓ	Ⓔ
20.	Ⓐ	Ⓑ	Ⓒ	Ⓓ	Ⓔ
21.	Ⓐ	Ⓑ	Ⓒ	Ⓓ	Ⓔ
22.	Ⓐ	Ⓑ	Ⓒ	Ⓓ	Ⓔ
23.	Ⓐ	Ⓑ	Ⓒ	Ⓓ	Ⓔ
24.	Ⓐ	Ⓑ	Ⓒ	Ⓓ	Ⓔ
25.	Ⓐ	Ⓑ	Ⓒ	Ⓓ	Ⓔ
26.	Ⓐ	Ⓑ	Ⓒ	Ⓓ	Ⓔ
27.	Ⓐ	Ⓑ	Ⓒ	Ⓓ	Ⓔ
28.	Ⓐ	Ⓑ	Ⓒ	Ⓓ	Ⓔ
29.	Ⓐ	Ⓑ	Ⓒ	Ⓓ	Ⓔ
30.	Ⓐ	Ⓑ	Ⓒ	Ⓓ	Ⓔ
31.	Ⓐ	Ⓑ	Ⓒ	Ⓓ	Ⓔ
32.	Ⓐ	Ⓑ	Ⓒ	Ⓓ	Ⓔ
33.	Ⓐ	Ⓑ	Ⓒ	Ⓓ	Ⓔ
34.	Ⓐ	Ⓑ	Ⓒ	Ⓓ	Ⓔ
35.	Ⓐ	Ⓑ	Ⓒ	Ⓓ	Ⓔ
36.	Ⓐ	Ⓑ	Ⓒ	Ⓓ	Ⓔ
37.	Ⓐ	Ⓑ	Ⓒ	Ⓓ	Ⓔ
38.	Ⓐ	Ⓑ	Ⓒ	Ⓓ	Ⓔ
39.	Ⓐ	Ⓑ	Ⓒ	Ⓓ	Ⓔ
40.	Ⓐ	Ⓑ	Ⓒ	Ⓓ	Ⓔ
41.	Ⓐ	Ⓑ	Ⓒ	Ⓓ	Ⓔ
42.	Ⓐ	Ⓑ	Ⓒ	Ⓓ	Ⓔ
43.	Ⓐ	Ⓑ	Ⓒ	Ⓓ	Ⓔ
44.	Ⓐ	Ⓑ	Ⓒ	Ⓓ	Ⓔ
45.	Ⓐ	Ⓑ	Ⓒ	Ⓓ	Ⓔ
46.	Ⓐ	Ⓑ	Ⓒ	Ⓓ	Ⓔ
47.	Ⓐ	Ⓑ	Ⓒ	Ⓓ	Ⓔ
48.	Ⓐ	Ⓑ	Ⓒ	Ⓓ	Ⓔ
49.	Ⓐ	Ⓑ	Ⓒ	Ⓓ	Ⓔ
50.	Ⓐ	Ⓑ	Ⓒ	Ⓓ	Ⓔ

Model Test 11 Answer Sheet

#							#					
1.	Ⓐ	Ⓑ	Ⓒ	Ⓓ	Ⓔ		26.	Ⓐ	Ⓑ	Ⓒ	Ⓓ	Ⓔ
2.	Ⓐ	Ⓑ	Ⓒ	Ⓓ	Ⓔ		27.	Ⓐ	Ⓑ	Ⓒ	Ⓓ	Ⓔ
3.	Ⓐ	Ⓑ	Ⓒ	Ⓓ	Ⓔ		28.	Ⓐ	Ⓑ	Ⓒ	Ⓓ	Ⓔ
4.	Ⓐ	Ⓑ	Ⓒ	Ⓓ	Ⓔ		29.	Ⓐ	Ⓑ	Ⓒ	Ⓓ	Ⓔ
5.	Ⓐ	Ⓑ	Ⓒ	Ⓓ	Ⓔ		30.	Ⓐ	Ⓑ	Ⓒ	Ⓓ	Ⓔ
6.	Ⓐ	Ⓑ	Ⓒ	Ⓓ	Ⓔ		31.	Ⓐ	Ⓑ	Ⓒ	Ⓓ	Ⓔ
7.	Ⓐ	Ⓑ	Ⓒ	Ⓓ	Ⓔ		32.	Ⓐ	Ⓑ	Ⓒ	Ⓓ	Ⓔ
8.	Ⓐ	Ⓑ	Ⓒ	Ⓓ	Ⓔ		33.	Ⓐ	Ⓑ	Ⓒ	Ⓓ	Ⓔ
9.	Ⓐ	Ⓑ	Ⓒ	Ⓓ	Ⓔ		34.	Ⓐ	Ⓑ	Ⓒ	Ⓓ	Ⓔ
10.	Ⓐ	Ⓑ	Ⓒ	Ⓓ	Ⓔ		35.	Ⓐ	Ⓑ	Ⓒ	Ⓓ	Ⓔ
11.	Ⓐ	Ⓑ	Ⓒ	Ⓓ	Ⓔ		36.	Ⓐ	Ⓑ	Ⓒ	Ⓓ	Ⓔ
12.	Ⓐ	Ⓑ	Ⓒ	Ⓓ	Ⓔ		37.	Ⓐ	Ⓑ	Ⓒ	Ⓓ	Ⓔ
13.	Ⓐ	Ⓑ	Ⓒ	Ⓓ	Ⓔ		38.	Ⓐ	Ⓑ	Ⓒ	Ⓓ	Ⓔ
14.	Ⓐ	Ⓑ	Ⓒ	Ⓓ	Ⓔ		39.	Ⓐ	Ⓑ	Ⓒ	Ⓓ	Ⓔ
15.	Ⓐ	Ⓑ	Ⓒ	Ⓓ	Ⓔ		40.	Ⓐ	Ⓑ	Ⓒ	Ⓓ	Ⓔ
16.	Ⓐ	Ⓑ	Ⓒ	Ⓓ	Ⓔ		41.	Ⓐ	Ⓑ	Ⓒ	Ⓓ	Ⓔ
17.	Ⓐ	Ⓑ	Ⓒ	Ⓓ	Ⓔ		42.	Ⓐ	Ⓑ	Ⓒ	Ⓓ	Ⓔ
18.	Ⓐ	Ⓑ	Ⓒ	Ⓓ	Ⓔ		43.	Ⓐ	Ⓑ	Ⓒ	Ⓓ	Ⓔ
19.	Ⓐ	Ⓑ	Ⓒ	Ⓓ	Ⓔ		44.	Ⓐ	Ⓑ	Ⓒ	Ⓓ	Ⓔ
20.	Ⓐ	Ⓑ	Ⓒ	Ⓓ	Ⓔ		45.	Ⓐ	Ⓑ	Ⓒ	Ⓓ	Ⓔ
21.	Ⓐ	Ⓑ	Ⓒ	Ⓓ	Ⓔ		46.	Ⓐ	Ⓑ	Ⓒ	Ⓓ	Ⓔ
22.	Ⓐ	Ⓑ	Ⓒ	Ⓓ	Ⓔ		47.	Ⓐ	Ⓑ	Ⓒ	Ⓓ	Ⓔ
23.	Ⓐ	Ⓑ	Ⓒ	Ⓓ	Ⓔ		48.	Ⓐ	Ⓑ	Ⓒ	Ⓓ	Ⓔ
24.	Ⓐ	Ⓑ	Ⓒ	Ⓓ	Ⓔ		49.	Ⓐ	Ⓑ	Ⓒ	Ⓓ	Ⓔ
25.	Ⓐ	Ⓑ	Ⓒ	Ⓓ	Ⓔ		50.	Ⓐ	Ⓑ	Ⓒ	Ⓓ	Ⓔ

Model Test 12 Answer Sheet

#							#					
1.	Ⓐ	Ⓑ	Ⓒ	Ⓓ	Ⓔ		26.	Ⓐ	Ⓑ	Ⓒ	Ⓓ	Ⓔ
2.	Ⓐ	Ⓑ	Ⓒ	Ⓓ	Ⓔ		27.	Ⓐ	Ⓑ	Ⓒ	Ⓓ	Ⓔ
3.	Ⓐ	Ⓑ	Ⓒ	Ⓓ	Ⓔ		28.	Ⓐ	Ⓑ	Ⓒ	Ⓓ	Ⓔ
4.	Ⓐ	Ⓑ	Ⓒ	Ⓓ	Ⓔ		29.	Ⓐ	Ⓑ	Ⓒ	Ⓓ	Ⓔ
5.	Ⓐ	Ⓑ	Ⓒ	Ⓓ	Ⓔ		30.	Ⓐ	Ⓑ	Ⓒ	Ⓓ	Ⓔ
6.	Ⓐ	Ⓑ	Ⓒ	Ⓓ	Ⓔ		31.	Ⓐ	Ⓑ	Ⓒ	Ⓓ	Ⓔ
7.	Ⓐ	Ⓑ	Ⓒ	Ⓓ	Ⓔ		32.	Ⓐ	Ⓑ	Ⓒ	Ⓓ	Ⓔ
8.	Ⓐ	Ⓑ	Ⓒ	Ⓓ	Ⓔ		33.	Ⓐ	Ⓑ	Ⓒ	Ⓓ	Ⓔ
9.	Ⓐ	Ⓑ	Ⓒ	Ⓓ	Ⓔ		34.	Ⓐ	Ⓑ	Ⓒ	Ⓓ	Ⓔ
10.	Ⓐ	Ⓑ	Ⓒ	Ⓓ	Ⓔ		35.	Ⓐ	Ⓑ	Ⓒ	Ⓓ	Ⓔ
11.	Ⓐ	Ⓑ	Ⓒ	Ⓓ	Ⓔ		36.	Ⓐ	Ⓑ	Ⓒ	Ⓓ	Ⓔ
12.	Ⓐ	Ⓑ	Ⓒ	Ⓓ	Ⓔ		37.	Ⓐ	Ⓑ	Ⓒ	Ⓓ	Ⓔ
13.	Ⓐ	Ⓑ	Ⓒ	Ⓓ	Ⓔ		38.	Ⓐ	Ⓑ	Ⓒ	Ⓓ	Ⓔ
14.	Ⓐ	Ⓑ	Ⓒ	Ⓓ	Ⓔ		39.	Ⓐ	Ⓑ	Ⓒ	Ⓓ	Ⓔ
15.	Ⓐ	Ⓑ	Ⓒ	Ⓓ	Ⓔ		40.	Ⓐ	Ⓑ	Ⓒ	Ⓓ	Ⓔ
16.	Ⓐ	Ⓑ	Ⓒ	Ⓓ	Ⓔ		41.	Ⓐ	Ⓑ	Ⓒ	Ⓓ	Ⓔ
17.	Ⓐ	Ⓑ	Ⓒ	Ⓓ	Ⓔ		42.	Ⓐ	Ⓑ	Ⓒ	Ⓓ	Ⓔ
18.	Ⓐ	Ⓑ	Ⓒ	Ⓓ	Ⓔ		43.	Ⓐ	Ⓑ	Ⓒ	Ⓓ	Ⓔ
19.	Ⓐ	Ⓑ	Ⓒ	Ⓓ	Ⓔ		44.	Ⓐ	Ⓑ	Ⓒ	Ⓓ	Ⓔ
20.	Ⓐ	Ⓑ	Ⓒ	Ⓓ	Ⓔ		45.	Ⓐ	Ⓑ	Ⓒ	Ⓓ	Ⓔ
21.	Ⓐ	Ⓑ	Ⓒ	Ⓓ	Ⓔ		46.	Ⓐ	Ⓑ	Ⓒ	Ⓓ	Ⓔ
22.	Ⓐ	Ⓑ	Ⓒ	Ⓓ	Ⓔ		47.	Ⓐ	Ⓑ	Ⓒ	Ⓓ	Ⓔ
23.	Ⓐ	Ⓑ	Ⓒ	Ⓓ	Ⓔ		48.	Ⓐ	Ⓑ	Ⓒ	Ⓓ	Ⓔ
24.	Ⓐ	Ⓑ	Ⓒ	Ⓓ	Ⓔ		49.	Ⓐ	Ⓑ	Ⓒ	Ⓓ	Ⓔ
25.	Ⓐ	Ⓑ	Ⓒ	Ⓓ	Ⓔ		50.	Ⓐ	Ⓑ	Ⓒ	Ⓓ	Ⓔ

Model Test 13 Answer Sheet

1.	Ⓐ Ⓑ Ⓒ Ⓓ Ⓔ	26. Ⓐ Ⓑ Ⓒ Ⓓ Ⓔ
2.	Ⓐ Ⓑ Ⓒ Ⓓ Ⓔ	27. Ⓐ Ⓑ Ⓒ Ⓓ Ⓔ
3.	Ⓐ Ⓑ Ⓒ Ⓓ Ⓔ	28. Ⓐ Ⓑ Ⓒ Ⓓ Ⓔ
4.	Ⓐ Ⓑ Ⓒ Ⓓ Ⓔ	29. Ⓐ Ⓑ Ⓒ Ⓓ Ⓔ
5.	Ⓐ Ⓑ Ⓒ Ⓓ Ⓔ	30. Ⓐ Ⓑ Ⓒ Ⓓ Ⓔ
6.	Ⓐ Ⓑ Ⓒ Ⓓ Ⓔ	31. Ⓐ Ⓑ Ⓒ Ⓓ Ⓔ
7.	Ⓐ Ⓑ Ⓒ Ⓓ Ⓔ	32. Ⓐ Ⓑ Ⓒ Ⓓ Ⓔ
8.	Ⓐ Ⓑ Ⓒ Ⓓ Ⓔ	33. Ⓐ Ⓑ Ⓒ Ⓓ Ⓔ
9.	Ⓐ Ⓑ Ⓒ Ⓓ Ⓔ	34. Ⓐ Ⓑ Ⓒ Ⓓ Ⓔ
10.	Ⓐ Ⓑ Ⓒ Ⓓ Ⓔ	35. Ⓐ Ⓑ Ⓒ Ⓓ Ⓔ
11.	Ⓐ Ⓑ Ⓒ Ⓓ Ⓔ	36. Ⓐ Ⓑ Ⓒ Ⓓ Ⓔ
12.	Ⓐ Ⓑ Ⓒ Ⓓ Ⓔ	37. Ⓐ Ⓑ Ⓒ Ⓓ Ⓔ
13.	Ⓐ Ⓑ Ⓒ Ⓓ Ⓔ	38. Ⓐ Ⓑ Ⓒ Ⓓ Ⓔ
14.	Ⓐ Ⓑ Ⓒ Ⓓ Ⓔ	39. Ⓐ Ⓑ Ⓒ Ⓓ Ⓔ
15.	Ⓐ Ⓑ Ⓒ Ⓓ Ⓔ	40. Ⓐ Ⓑ Ⓒ Ⓓ Ⓔ
16.	Ⓐ Ⓑ Ⓒ Ⓓ Ⓔ	41. Ⓐ Ⓑ Ⓒ Ⓓ Ⓔ
17.	Ⓐ Ⓑ Ⓒ Ⓓ Ⓔ	42. Ⓐ Ⓑ Ⓒ Ⓓ Ⓔ
18.	Ⓐ Ⓑ Ⓒ Ⓓ Ⓔ	43. Ⓐ Ⓑ Ⓒ Ⓓ Ⓔ
19.	Ⓐ Ⓑ Ⓒ Ⓓ Ⓔ	44. Ⓐ Ⓑ Ⓒ Ⓓ Ⓔ
20.	Ⓐ Ⓑ Ⓒ Ⓓ Ⓔ	45. Ⓐ Ⓑ Ⓒ Ⓓ Ⓔ
21.	Ⓐ Ⓑ Ⓒ Ⓓ Ⓔ	46. Ⓐ Ⓑ Ⓒ Ⓓ Ⓔ
22.	Ⓐ Ⓑ Ⓒ Ⓓ Ⓔ	47. Ⓐ Ⓑ Ⓒ Ⓓ Ⓔ
23.	Ⓐ Ⓑ Ⓒ Ⓓ Ⓔ	48. Ⓐ Ⓑ Ⓒ Ⓓ Ⓔ
24.	Ⓐ Ⓑ Ⓒ Ⓓ Ⓔ	49. Ⓐ Ⓑ Ⓒ Ⓓ Ⓔ
25.	Ⓐ Ⓑ Ⓒ Ⓓ Ⓔ	50. Ⓐ Ⓑ Ⓒ Ⓓ Ⓔ

Model Test 14 Answer Sheet

1.	Ⓐ Ⓑ Ⓒ Ⓓ Ⓔ	26. Ⓐ Ⓑ Ⓒ Ⓓ Ⓔ
2.	Ⓐ Ⓑ Ⓒ Ⓓ Ⓔ	27. Ⓐ Ⓑ Ⓒ Ⓓ Ⓔ
3.	Ⓐ Ⓑ Ⓒ Ⓓ Ⓔ	28. Ⓐ Ⓑ Ⓒ Ⓓ Ⓔ
4.	Ⓐ Ⓑ Ⓒ Ⓓ Ⓔ	29. Ⓐ Ⓑ Ⓒ Ⓓ Ⓔ
5.	Ⓐ Ⓑ Ⓒ Ⓓ Ⓔ	30. Ⓐ Ⓑ Ⓒ Ⓓ Ⓔ
6.	Ⓐ Ⓑ Ⓒ Ⓓ Ⓔ	31. Ⓐ Ⓑ Ⓒ Ⓓ Ⓔ
7.	Ⓐ Ⓑ Ⓒ Ⓓ Ⓔ	32. Ⓐ Ⓑ Ⓒ Ⓓ Ⓔ
8.	Ⓐ Ⓑ Ⓒ Ⓓ Ⓔ	33. Ⓐ Ⓑ Ⓒ Ⓓ Ⓔ
9.	Ⓐ Ⓑ Ⓒ Ⓓ Ⓔ	34. Ⓐ Ⓑ Ⓒ Ⓓ Ⓔ
10.	Ⓐ Ⓑ Ⓒ Ⓓ Ⓔ	35. Ⓐ Ⓑ Ⓒ Ⓓ Ⓔ
11.	Ⓐ Ⓑ Ⓒ Ⓓ Ⓔ	36. Ⓐ Ⓑ Ⓒ Ⓓ Ⓔ
12.	Ⓐ Ⓑ Ⓒ Ⓓ Ⓔ	37. Ⓐ Ⓑ Ⓒ Ⓓ Ⓔ
13.	Ⓐ Ⓑ Ⓒ Ⓓ Ⓔ	38. Ⓐ Ⓑ Ⓒ Ⓓ Ⓔ
14.	Ⓐ Ⓑ Ⓒ Ⓓ Ⓔ	39. Ⓐ Ⓑ Ⓒ Ⓓ Ⓔ
15.	Ⓐ Ⓑ Ⓒ Ⓓ Ⓔ	40. Ⓐ Ⓑ Ⓒ Ⓓ Ⓔ
16.	Ⓐ Ⓑ Ⓒ Ⓓ Ⓔ	41. Ⓐ Ⓑ Ⓒ Ⓓ Ⓔ
17.	Ⓐ Ⓑ Ⓒ Ⓓ Ⓔ	42. Ⓐ Ⓑ Ⓒ Ⓓ Ⓔ
18.	Ⓐ Ⓑ Ⓒ Ⓓ Ⓔ	43. Ⓐ Ⓑ Ⓒ Ⓓ Ⓔ
19.	Ⓐ Ⓑ Ⓒ Ⓓ Ⓔ	44. Ⓐ Ⓑ Ⓒ Ⓓ Ⓔ
20.	Ⓐ Ⓑ Ⓒ Ⓓ Ⓔ	45. Ⓐ Ⓑ Ⓒ Ⓓ Ⓔ
21.	Ⓐ Ⓑ Ⓒ Ⓓ Ⓔ	46. Ⓐ Ⓑ Ⓒ Ⓓ Ⓔ
22.	Ⓐ Ⓑ Ⓒ Ⓓ Ⓔ	47. Ⓐ Ⓑ Ⓒ Ⓓ Ⓔ
23.	Ⓐ Ⓑ Ⓒ Ⓓ Ⓔ	48. Ⓐ Ⓑ Ⓒ Ⓓ Ⓔ
24.	Ⓐ Ⓑ Ⓒ Ⓓ Ⓔ	49. Ⓐ Ⓑ Ⓒ Ⓓ Ⓔ
25.	Ⓐ Ⓑ Ⓒ Ⓓ Ⓔ	50. Ⓐ Ⓑ Ⓒ Ⓓ Ⓔ

Model Test 15 Answer Sheet

1.	Ⓐ	Ⓑ	Ⓒ	Ⓓ	Ⓔ		26.	Ⓐ	Ⓑ	Ⓒ	Ⓓ	Ⓔ
2.	Ⓐ	Ⓑ	Ⓒ	Ⓓ	Ⓔ		27.	Ⓐ	Ⓑ	Ⓒ	Ⓓ	Ⓔ
3.	Ⓐ	Ⓑ	Ⓒ	Ⓓ	Ⓔ		28.	Ⓐ	Ⓑ	Ⓒ	Ⓓ	Ⓔ
4.	Ⓐ	Ⓑ	Ⓒ	Ⓓ	Ⓔ		29.	Ⓐ	Ⓑ	Ⓒ	Ⓓ	Ⓔ
5.	Ⓐ	Ⓑ	Ⓒ	Ⓓ	Ⓔ		30.	Ⓐ	Ⓑ	Ⓒ	Ⓓ	Ⓔ
6.	Ⓐ	Ⓑ	Ⓒ	Ⓓ	Ⓔ		31.	Ⓐ	Ⓑ	Ⓒ	Ⓓ	Ⓔ
7.	Ⓐ	Ⓑ	Ⓒ	Ⓓ	Ⓔ		32.	Ⓐ	Ⓑ	Ⓒ	Ⓓ	Ⓔ
8.	Ⓐ	Ⓑ	Ⓒ	Ⓓ	Ⓔ		33.	Ⓐ	Ⓑ	Ⓒ	Ⓓ	Ⓔ
9.	Ⓐ	Ⓑ	Ⓒ	Ⓓ	Ⓔ		34.	Ⓐ	Ⓑ	Ⓒ	Ⓓ	Ⓔ
10.	Ⓐ	Ⓑ	Ⓒ	Ⓓ	Ⓔ		35.	Ⓐ	Ⓑ	Ⓒ	Ⓓ	Ⓔ
11.	Ⓐ	Ⓑ	Ⓒ	Ⓓ	Ⓔ		36.	Ⓐ	Ⓑ	Ⓒ	Ⓓ	Ⓔ
12.	Ⓐ	Ⓑ	Ⓒ	Ⓓ	Ⓔ		37.	Ⓐ	Ⓑ	Ⓒ	Ⓓ	Ⓔ
13.	Ⓐ	Ⓑ	Ⓒ	Ⓓ	Ⓔ		38.	Ⓐ	Ⓑ	Ⓒ	Ⓓ	Ⓔ
14.	Ⓐ	Ⓑ	Ⓒ	Ⓓ	Ⓔ		39.	Ⓐ	Ⓑ	Ⓒ	Ⓓ	Ⓔ
15.	Ⓐ	Ⓑ	Ⓒ	Ⓓ	Ⓔ		40.	Ⓐ	Ⓑ	Ⓒ	Ⓓ	Ⓔ
16.	Ⓐ	Ⓑ	Ⓒ	Ⓓ	Ⓔ		41.	Ⓐ	Ⓑ	Ⓒ	Ⓓ	Ⓔ
17.	Ⓐ	Ⓑ	Ⓒ	Ⓓ	Ⓔ		42.	Ⓐ	Ⓑ	Ⓒ	Ⓓ	Ⓔ
18.	Ⓐ	Ⓑ	Ⓒ	Ⓓ	Ⓔ		43.	Ⓐ	Ⓑ	Ⓒ	Ⓓ	Ⓔ
19.	Ⓐ	Ⓑ	Ⓒ	Ⓓ	Ⓔ		44.	Ⓐ	Ⓑ	Ⓒ	Ⓓ	Ⓔ
20.	Ⓐ	Ⓑ	Ⓒ	Ⓓ	Ⓔ		45.	Ⓐ	Ⓑ	Ⓒ	Ⓓ	Ⓔ
21.	Ⓐ	Ⓑ	Ⓒ	Ⓓ	Ⓔ		46.	Ⓐ	Ⓑ	Ⓒ	Ⓓ	Ⓔ
22.	Ⓐ	Ⓑ	Ⓒ	Ⓓ	Ⓔ		47.	Ⓐ	Ⓑ	Ⓒ	Ⓓ	Ⓔ
23.	Ⓐ	Ⓑ	Ⓒ	Ⓓ	Ⓔ		48.	Ⓐ	Ⓑ	Ⓒ	Ⓓ	Ⓔ
24.	Ⓐ	Ⓑ	Ⓒ	Ⓓ	Ⓔ		49.	Ⓐ	Ⓑ	Ⓒ	Ⓓ	Ⓔ
25.	Ⓐ	Ⓑ	Ⓒ	Ⓓ	Ⓔ		50.	Ⓐ	Ⓑ	Ⓒ	Ⓓ	Ⓔ

"Shorten 40 hours of college preparatory precalculus study to an easy 4 hours…"

- The usage of the TI 83 – TI 84 family of graphing calculators particularly in the context of Algebra, Pre-Calculus and SAT and IB Mathematics with over 400 questions carefully designed and fully solved questions;
- A detailed additional chapter on TI 89 with a special focus on the algebra menu and the graphing facilities of this device;
- 10 original and fully solved sample tests 5 or Level 1 and 5 for Level 2.

This book is intended to help high school students who are bound to take either or both of the SAT Math Level 1 and Level 2 tests. Being one of a kind, this book is devoted to the usage of the TI 83 – TI 84 family of graphing calculators particularly in the context of Algebra, Pre-Calculus and SAT and IB Mathematics with over 400 questions carefully designed and fully solved questions. The method proposed in this book has been developed through 5 years' experience; has been proven to work and has created a success story each and every time, having helped hundreds of students who are currently attending the top 50 universities in the USA including many Ivy League schools.

The main advantage of the approach suggested in this book is that, one can solve, any equation or inequality with the TI, whether it is algebraic, trigonometric, exponential, logarithmic, polynomial or one that involves absolute values, without needing to know the related topic in depth and having to perform tedious steps One can solve all types of equations and inequalities very easily and in a very similar way just needing to learn a few very easy to remember techniques.

But there are still more to what can be done with the TI; find period, frequency, amplitude, offset, axis of wave of a periodic function, find the maxima minima and zeros as well as the domains and ranges of all types of functions; perform any operations on complex numbers, carry out any computation involving sequences and series, perform matrix algebra, solve a system of equations for any number of unknowns and even write small programs to ease your life. More than 20 of the 50 questions in the SAT Mathematics Subject Tests are based on the topics given above and this is why this book upgrades the SAT Mathematics subject test scores by at least 200 points.

Topics covered are:

- EQUATIONS: Polynomial, Algebraic, Absolute Value, Exponential and Logarithmic, Trigonometric, Inverse Trigonometric
- INEQUALITIES: Polynomial, Algebraic and Absolute Value, Trigonometric
- FUNCTIONS: Maxima and Minima, Domains and Ranges, Evenness and Oddness, Graphs of Trigonometric Functions, the Greatest Integer Function
- BASIC CALCULUS: Zeros, Holes, Limits, Continuity, Horizontal and Vertical Asymptotes
- CONIC SECTIONS: Circle, Ellipse, Parabola and Hyperbola
- LINEAR ALGEBRA: System of Linear Equations, Matrices and Determinants
- MISCELLANEOUS: Parametric and Polar Graphing; Complex Numbers; Permutations and Combinations, Computer Programs, Sequences and Series, Statistics, and more…

Complete Prep for the SAT Math Subject Tests
Level 1 and Level 2 with 10 Fully Solved Sample Tests

First Edition

RUŞEN MEYLANİ, M.S.

Complete Prep for the SAT* Math Subject Tests Level 1 & Level 2
up-to-date and true-to-life

with 10 FULLY SOLVED SAMPLE TESTS

- The only complete prep for the SAT Math Level 1 and Level 2 Subject Tests
- 10 Original Sample Tests, 5 for Level 1 and 5 for Level 2
- Every topic that has ever appeared or is likely to appear on SAT Math Subject Tests, covered in detail and accurately
- Hundreds of fully solved examples and more than a thousand exercises
- It is highly recommended that this book be used along with the other titles of the same author for a perfect score

RUSH
Publications and Educational Consultancy

The only complete prep for the SAT Math Level 1 and Level 2 Subject Tests

The book covers every topic that has ever appeared on both tests in detail and accurately. Moreover there are 10 sample tests, 5 from each level.

The topics covered are: Basic algebra; number theory; equations; inequalities; unit conversions; logic; arithmetic, geometric and harmonic means; basic computer programs; operations; functions; evenness and oddness; basic functions; transformations; advanced graphing; symmetries; rotations; linear functions; quadratic functions; polynomial functions; trigonometry; exponentials and logarithms; limits, continuity and asymptotes; the greatest integer function; absolute values; conic sections; complex numbers; parametric and polar coordinates; variation; locus; plane and three dimensional geometry; inscribed figures (two or three dimensional); rotations in three dimensions; permutations, combinations and probability; binomials; sequences and series; statistics; matrices and determinants; three dimensional coordinate geometry; vectors; data analysis (graphs and tables) and all kinds of word problems.

About the Author

Ruşen* Meylani is the co – founder of **RUSH** Publications and Educational Consultancy, LLC. Born on the 2nd of August, 1972 he shows all the leadership characteristics of a Leo. He was awarded many times in mathematics since the age of 14 and he holds a Master of Science degree in Electrical and Electronics Engineering having written tens of papers in this field. When he was a graduate student (between 1995 and 1997), he invented three methods that are currently being used with sophisticated printing devices, the last generation white goods, and, the state of the art quality control systems. He worked in the field of Information Technology where he was the leader of a team that built the wide area network of a major supermarket chain in Europe.

In 1998 he decided to build a career in education listening to the sound of his heart and that was it. He created a method that shortens 40 hours of mathematics to 4 hours gaining the attraction of Eisenhower National Clearinghouse funded by the US Department of Education. In 2004 and 2005 he gave several conferences in the United States on this particular method. In the mean time he published three SAT II Mathematics books that became bestselling among their peers in Amazon just in a few months. He has 15 other books that will have all been published by mid 2007. He is a researcher, educator and academician having taught at several distinguished institutions at the K9 – 12, undergraduate and graduate levels, all in Europe. He has dedicated his life to creating easy to teach and easy to learn methods in mathematics. He considers himself as a "gifted loony" as he claims that one day it will be possible to shorten the learning time to less than one tenth of the usual. When people ask him about how it will be done, he borrows Albert Einstein's words: "If at first the idea is not absurd then there is no hope for it."

Ruşen Meylani's motto is somewhat similar to Robert Kennedy's: "Some men see things as they are and say 'Why?' I see the things that never were and say 'Why not!' "

---

*ş is pronounced as "sh" like in "Ash Wednesday" (a poem by T.S. Elliot)

 **collegeboard.com**

SAT Registration & Scores

SAT Registration

RUSEN MEYLANI

Update SAT Personal Info

Date	Test	Score	Percentile*	Status
10/2004	**SAT II**			Test Completed
	Math Level 1 w/Calculator	800	99	
	Math Level 2 w/Calculator	800	90	
10/2002	**SAT II**			Test Completed
	Math Level 1 w/Calculator	800	99	
	Physics	800	93	
05/2002	**SAT II**			Test Completed
	Math Level 1 w/Calculator	800	99	
	Math Level 2 w/Calculator	800	90	
	Physics	800	93	
06/2000	**SAT II**			Test Completed
	Math Level 1 w/Calculator	800	99	
	Math Level 2 w/Calculator	800	90	
	Physics	800	93	
05/2000	**SAT II**			Test Completed
	Math Level 1 w/Calculator	800	99	
	Math Level 2 w/Calculator	800	90	

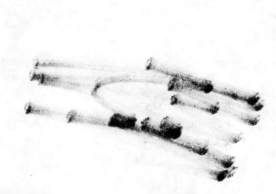